On the Keeping of Bees

By
John M Whitaker

Northern Bee Books

On the Keeping of Bees

© John M Whitaker

ISBN 978-1-912271-48-1

Published by Northern Bee Books, 2019
Scout Bottom Farm
Mytholmroyd
Hebden Bridge HX7 5JS (UK)

Design and artwork by DM Design and Print

Dedicated to the memory of Eileen Ramsden,
died 22nd February 2010

Contents

Chapter 1

Why keep honeybees?

On the day I finished the first edition of this book, before breakfast, I took the dogs for a walk. It was early September and the sun was just rising above the horizon, struggling and failing to burn through the cloud cover and so it was still gloomy and there was an autumnal chill in the still air. As usual I strolled across the paddock to see the bees. The older wiser dog sat down fifty metres from the apiary, prepared to wait for my return. The puppy bounced along beside me. From twenty metres away I could hear the distinct hum of the hives and leaning over the gate of the apiary I saw the bees pouring from the hives and some were already returning with their first loads of nectar and pollen of the day. There to see was a wonder of nature, a mystery and, for me, a source of endless fascination.

If anyone ever reads this book – well and good. But the main beneficiary from it, I believe has been and will be my own beekeeping. There is nothing like writing things down to clarify one's own thoughts.

The book reflects my beekeeping philosophy and methods. I don't set myself up as a paragon of beekeeping or of anything else come to that. I have a reasonable idea of what I am trying to achieve but like most honest beekeepers I don't hesitate to admit that my achievements fall far short of my aims. For a number of years I've run beekeeping courses for beginners and the contents of the course have formed the backbone of what I'm writing here.

Also, and this is in retrospect, this book contains the knowledge that a new beekeeper would require to prepare for the BBKA Basic Husbandry Examination.

I could be described as being very parochial in my approach. For twenty-four years I kept my bees in the Vale of York, though just recently my wife and I have moved, with my bees of course, to the beautiful county of Herefordshire in order to be able to watch our grandchildren growing up. But the vast majority of my experience remains Yorkshire centric. Beekeeping is, by its very nature, parochial. But this is not to say that what is said about beekeeping in Yorkshire (or Herefordshire) does not have validity or interest elsewhere.

There are two themes that will run through these pages

1) I believe that a knowledge of the natural history of the honeybee, both of the individual insect and the superorganism, is a prerequisite for good beekeeping

2) I believe that the first priority of beekeeping husbandry is to do everything that is necessary to keep the bees healthy.

This is not written telling you how to produce lots of honey. If you keep your bees healthy and strong, if the weather is good, if you have reasonable queens and if you live in an area with a variety of forage, then you will get honey whether you want it or not. A few years ago an elderly beekeeper, who had been a member of our association for several decades, died. He was a widower, had lived a full life and had kept bees for many, many years and his vast experience was freely available to any new beekeeper in the association. A real gent. In fact he died while tending his bees, but that is bye the bye. The family asked some of our association members to help sort out the beekeeping side of his life and when they went through his house in every cupboard, under beds, piled up in spare rooms were 15kg buckets of top class honey, about 80 of them altogether, well over a metric tonne of honey. He was a gentleman beekeeper to whom the bees were more important than the honey.

When I took up beekeeping at the age of 45 I believe I found my true niche in life. As a boy, my cousin and I played in my uncle's garden where there were three WBC hives at the far end beneath a pair of plum trees. To my shame I displayed no curiosity in the bees and to my uncle's shame he made no effort to introduce me to them. We were just told to keep away from them. And at home our garden backed on to the apiary of the local school where they kept bees and so occasionally a swarm would appear on our apple tree. But again, my mother insisted that we children remain indoors until the teacher had arrived to remove the swarm. Yet, despite these unpromising passing acquaintances with bees, now in midlife, married and a father, I knew I wanted to keep bees. The first step was to attend an evening class run by Bill Bell, a wellknown York beekeeper. Bill Bell was an excellent beekeeper and a great raconteur. For the first years I followed his methodology but as I gained experience I went my own way, and now I wonder how much of his teaching he would recognise in my beekeeping.

Diversity of approach is one of the joys of beekeeping. There are countless wrong ways to keep bees but also countless perfectly proper ways to keep bees. When we set out to keep bees there are a multitude of questions that need to be asked

What type of hive should we use?

What type of frames should we use?

What type of frame spacing should we use?

What type of queen excluder should we use?

Should we use top bee space or bottom bee space?

How many hives should we keep?

How do we sell the honey?

Should we show our honey?

And there are many, many more questions. All the questions listed above relate to options which require a choice from the beekeeper. Unlike schoolboy mathematics, there is no one correct answer to any of these questions. In the following pages these questions and many others will be addressed, but not necessarily answered in a specific way. Our beekeeping is also limited and constricted by the conditions that are imposed upon us by the area in which we live, the topography, the local flora, the weather and climate and the time and expense that we are willing to invest in beekeeping. Each of these factors results in furthering the uniqueness of each beekeeper as to their approach to beekeeping

So we can safely say that no two beekeepers approach their beekeeping in the same way. This diversity of approach leads to many enjoyable hours in the local pub or leaning on the apiary gate engaged in lively, often heated, discussion with fellow beekeepers. And beekeepers are themselves a diverse group, railwaymen, teachers, joiners, farmers, doctors, nurses, men of the cloth, technicians, chemists, cooks, financial advisors, members of the armed forces and members of HMP system. They are male and female from working class and professional backgrounds, young and old, loners and team workers. Beekeepers in the UK are not generally professionals, but rather most are hobbyists. It is possible to make a living from beekeeping but only a few beekeepers in this country do so. But central to this diversity there is a respect for each other based on this common fascination with this little insect, the honeybee.

Beekeeping Associations

I believe that beekeeping is best carried out within an association and every part of England is served by a beekeeping association. There are good reasons for this. Beekeeping is a knowledge rich craft and the diversity means that there is never a time when you are not able to learn from others. As a new beekeeper the advice and guidance that is received from those more experienced is absolutely invaluable. And beekeeping is not static. The bees can be threatened by disease and pests and as risks wax and wane so the best practice changes over the years. For example, beekeepers who did not adapt to the threat of varroa, lost their bees. Associations are an active channel for communicating best practice to their members. And then there is the society and friendship. In Yorkshire there are 21 area associations and these are affiliated to the Yorkshire Beekeepers Association, which is itself affiliated to the British Beekeepers Association. Other counties have a similar structure. The majority of activity is carried out at the local area, but events are also organised at the county and national level. The BBKA represents beekeepers at the governmental level, both within the UK and Europe. Membership of BBKA includes the provision of third party insurance and the BBKA manage an examination system, designed to raise the standards of beekeeping among its members.

The government also supports beekeeping, and in doing so tacitly acknowledges the value that honeybees give to agriculture, horticulture and to the environment. Throughout England there is a network of regional bee inspectors and seasonal bee inspectors employed by FERA, a department of DEFRA. The bee inspectors have a number of duties, but probably the one with the greatest profile is the monitoring of bees for disease. There are several diseases and pests that are by law notifiable to the bee inspectorate and when they are discovered the bee inspectors take the leading and vital role in eradicating the outbreak and ensuring that it is confined. In addition, they play a role in educating beekeepers about honeybee diseases. The government also fund research into honeybees, usually through universities, but also at the Rothmanstead research station and the National Bee Unit in York.

Yorkshire beekeeper in his Yorkshire apiary

An association apiary meeting

As implied in the previous paragraph beekeeping adds to the wealth of the country. Most crops that have colourful blossom, such as oil seed rape, borage, apples and pears, require insects to pollinate them. Honeybees are not the only insects that perform this role. Bumble bees and other social and solitary bees are equally important. Honeybees, however, are crucial pollinators in the early spring when honeybee populations are already established but other pollinators are just beginning their annual life cycle. The annual value of honeybees to the UK economy, that is the value of the honey, the wax and pollination of crops is estimated to be approximately £160 million. When it is also considered that there are approximately 200,000 colonies of honey bees kept by beekeepers in the UK, this would suggest that each colony contributes annually approximately £800. However, beekeepers need to accept that this value is not, and will never be, realised by the beekeepers themselves. Honeybees also pollinate wild flora and so make a contribution to maintaining our natural biodiversity of flora. The coming of the varroa mite has significantly reduced the number of feral colonies and so agriculture and the environment must depend to an even greater extent on the honeybees kept by beekeepers.

This by itself is a good reason to keep bees. But if you are not willing to commit to beekeeping for purely altruistic reasons, there are plenty of more selfish reasons to keep bees. You will produce honey and wax. The amount of honey per colony you produce will vary greatly from just a few pounds to over a hundredweight and depends on the weather throughout the summer, the prolificacy of the queen, the area in which you keep your bees and your skill as a beekeeper. The honey you will produce is not the same as the honey you buy in the supermarket – there is a delicacy of taste that you only get from local honey. And every batch of honey that you produce differs from all others.

But most important to me is the connection that beekeeping generates between the beekeeper and the natural world. We live in a world where our environment is almost totally created by our fellow man. We live in houses where we take no end of care to exclude animals, bacteria, viruses, we travel in manmade machines, and we work in sterile offices or factories using manmade machines. It is easy to accept this. It is comfortable and safe. But this divorce from the natural world truncates and reduces our own natures. We sense that we are being controlled by the pressures of a modern civilisation that make us less than we should be. Beekeeping allows us, in possibly just a small way, to reacquaint ourselves with the natural world about us. Its not just that we are learning to manage and study a truly wild creature, but we are forced to be aware of the seasons and the relationships between the fauna and flora that create the environment in which our honeybees and ourselves struggle to survive.

A colony of honeybees is without doubt a remarkably complex entity. Once you take up beekeeping you are embarking on a lifelong quest to try to understand its behaviour. The fascination with the honeybees never seems to die. No matter how many times a beekeeper opens up a hive there is a sense of wonder at the beauty, intricacy and industry of what you are observing.

Apiaries

Honeybees can be kept almost anywhere below a height of 300m in the UK, but some places are better than others. Contrary to one's intuition, the countryside is not necessarily the best place for honeybees. A significant proportion of the land in the countryside is devoted to arable farming, and the vast fields of cereal crops, wheat and barley, are a desert as far as the honeybee is concerned. Some flowering crops such as oil seed rape, field beans and borage can provide a large source of nectar for a short period of time. Improved grassland is also not helpful to honeybees. The application of nitrogen fertilisers has reduced the clover and other herbs to negligible amounts. So honeybees in the countryside, for most of the year, must exploit hedges, field margins, woods, gardens and riversides. Using google maps to view the countryside from the air you can see that in arable farming areas a remarkably small proportion of the land area falls into these categories. I keep bees in such an arable area but am fortunate to have small woods, a riverside and unimproved meadows close by. Nevertheless, there is usually a gap of four weeks from the middle of June to the middle of July when there is very little forage. On the other hand, urban, semi-urban or industrial margins can provide a continuous and varied source of forage. Urban areas also tend to be warmer than the countryside and especially during the early spring this can give the bees located in towns and cities an advantage.

It is undoubtedly true that there are good bee areas. These are areas where the bees seem to prosper better than others, despite a similarity of flora. Maybe this has to do with the topography which produces a more favourable microclimate or to the underlining geology. My best site in Yorkshire was on sandstone but I don't know whether that had a bearing.

Having said all that, the lack of the ideal site for an apiary should not be a discouragement. Bees, with sensible management, can prosper in most places. What is important is that the opportunities available are used effectively and the number of colonies kept at an apiary is commensurate with the foraging opportunities. Besides forage, there are other factors that should be taken into account when choosing an apiary site.

a)   The apiary should be sheltered from the north and open to sunshine from the south.

b)   There should be a source of water within 50m

c)   The hives should be positioned so that they do not become a nuisance to neighbours or passers by. If an apiary is being established in a populated area, then it should be surrounded by a 2m high fence. There should not be a line of sight to the hives from a public area of less than 20m.

d)   There needs to be vehicular access to within 20m or so of the hives. Hive boxes when they are full of honey are heavy.

e)   If establishing an apiary in the countryside, try to make the site secure, situating the apiary beyond a locked gate. It's sad that this sort of thing needs to be said, but it reflects the times we live in.

The number of hives that can be kept in one location has got a limit, and that limit depends upon the nature of the flora in the surrounding countryside. Once the colonies start competing against each other then a law of diminishing returns comes into play. In general, I limit the number of colonies in an apiary to eight. Besides, for management reasons, it is a good idea to have more than one apiary.

Over the years I've visited quite a few apiaries. Often they are hidden away in beautiful, peaceful, hidden places and it is a joy to spend a couple of hours there on a summer's day. The good apiary sites get passed on from beekeeper to beekeeper down the generations. Apiaries are much less common than they were 50 years ago, but, in my experience, this is not because of the lack of possible sites. In general farmers and landowners are happy to give access to suitable areas – it's often just a matter of asking.

The Sting

Honeybees do sting, and anyone who takes up beekeeping must expect to be stung. A bee sting hurts and should be avoided. Over the years there are few if any parts of my body that haven't been stung. But having said that it has never put me off and as often as not a sting is the result of my own carelessness and my own mistakes. Usually the bees can be opened for inspection without the beekeeper being stung. Some colonies are tetchier than others and colonies which are excessively aggressive need to be requeened. There are also times to avoid opening the hive, such as in thundery weather, in cold or windy weather, late in the evening, when a nectar flow has just finished or late in the season. One quickly becomes aware that the bees are unhappy as the pitch of the buzzing increases. In such circumstances the prudent thing is to close the hive and leave them to another day. But on a sunny spring or summer day, in the middle of the day, hives can be opened and the bees hardly seem to be aware that you are there. Good beekeeping practice also helps. Behaviour about the hives needs to be calm and controlled, with no sudden movements. The frames should not be knocked or dropped and care must be taken not to squash or distress the bees in any way. Smoke can help to subdue bees. This is an area where there can be contention and I'll return to it later. When a bee stings, pheromones are left at the site of the sting and these raise the level of aggression.

The sting definitely hurts, but not for long – maybe a couple of minutes. No matter how long I've kept bees it still hurts. An experienced beekeeper will confine himself or herself, in extremes, to the odd rude word, apologising, if in company, for the use of a technical beekeeping expression, and then get on with the job. When I started beekeeping there would be an itchy swelling at the site of the sting that would last 24 to 36 hours. This now rarely happens.

Stings on the hands and arms are of little consequence, but stings on the tender areas of the face, the lips, nose and eyes are not pleasant. For this reason it is prudent to always wear a veil. Most beekeepers now wear all in one suits, combining a veil with overalls. It is sensible to wear wellingtons to protect the ankles, which, for some reason, tend to be targeted by the bees. Gloves are another contentious issue. Thick leather gloves are widely used and do give a sense of security for the new beekeeper. However, there are arguments against using leather gloves. There is a loss of sensitivity in the handling of the bees and leather gloves are difficult to clean and can become a vector for passing pathogens from one colony to another, or even from one apiary to another. My preference is to use disposable nitrile gloves, along with a pair of gauntlets to protect the wrists.

Chapter 2
The natural history of the honeybee

The art of beekeeping is based upon an understanding of the natural history of the honeybee. One of the pleasures and challenges of beekeeping is that you never feel that your knowledge is complete, and it certainly won't be after reading this chapter which is intended to give just the basic and essential knowledge of honeybee biology that a beekeeper needs. No matter how long you keep bees you can always learn something new and the bees are forever displaying aspects of behaviour that it is difficult to understand. It is no surprise that for hundreds of years honeybees have been an object of curiosity, attracting the attention of amateur and professional researchers alike.

A Brief History of the understanding of the Honeybee

Aristotle wrote extensively about honeybes in Greece 2300 years ago. His writing was a strange mixture of remarkable insight, observation and gross error, but such was his reputation that little of what he wrote was challenged for almost 2000 years. Rome also produced writers who interested themselves in honeybees, including Virgil, Pliny the Elder and Columella. Columella was a practising beekeeper and his book, De Re Rustica is the best attempt in the ancient world at presenting the subject in a methodical way. One of the main errors of the ancients was the belief that a colony of honeybees is led by a king bee. This was finally corrected by Rev Charles Butler in his book, Feminine Monarchie, published in 1609. That he was writing following the reign of Queen Elizabeth the first probably created a more sympathetic intellectual climate for such a revaluation. The invention of the microscope and the spirit of curiosity engendered by the European Renaissance led to an accelerating understanding of the honeybee. Developments came from all parts of Europe. Anton Janscha (1734 – 1773), working in Vienna, documented how the swarm and queen mating occurred. The Swiss Francis Huber (1750 – 1831), though blind, devoted his life to furthering our understanding of wax production and comb building. The Rev John Dzierzon, a Polish priest, (1811 – 1906), was the first to realise that drones developed from unfertilised eggs. Professor Karl Von Frisch (1886 – 1982), working in Germany, interpreted the waggle dance in the 1940's, despite World War II raging.

The profit that the honeybees generate and the fact that they are enclosed in a hive that can be opened, manipulated and observed, invited experimentation and research. In 1923 a research centre was established at Rothamsted, England and this is still operating today. In addition the National Bee Unit in York has an experimental facility, and the university of Sussex is actively involved in research into Honeybees. In the last few decades several researchers in the United States have been taking a leading role in taking our knowledge forward, particularly Prof Thomas Seeley at Cornell University. Libraries of books and journals about honeybees and beekeepers have been written and continue to be produced, far more than about all other insects put together.

How does all this affect us, hobby beekeepers with one, two, ten, twenty hives at the end of the garden. It is simple. The better our understanding and knowledge of honeybees, the better we will be as beekeepers.

The Evolution of the Honeybee

In evolutionary timescales, the honeybee is a fairly recent development. The earth was created 4500 million years ago and the first life forms based on DNA are thought to have occurred 3500 million years ago. The first rocks that contain fossil evidence of insects are dated as being about 400 million years old. Honeybees seem to have evolved in a 10 million year period between 40 and 30 million years ago and in the last 30 million years have reached a stasis.

To illustrate the timescales more graphically, consider the history of the earth to date as a 24 hour day, with the earth being created at one second after midnight.

05:00 The appearance of the first primitive living creatures that were based on DNA, and so able to replicate.

21:20 The first vertebrates appear.

21:50 The first insects appear

23:20 Flowering plants begin to appear.

23:40 The extinction of the dinosaurs.

23:50 The first honeybees appear.

23:58 The first hominids appear

Like human beings, honeybees are recent arrivals, and like us, honeybees are sophisticated animals, not just able to exist in a specific niche, but able to adapt and create their own environment and therefore thrive over large parts of the earth, from the temperate regions into the tropics.

The honeybee evolved into its current form in Africa, and there are still a number of subspecies that inhabit the different regions of Africa. The honeybee then migrated north into Europe, and a number of different subspecies evolved in these more temperate areas. During the ice age the range of the European honeybee was restricted to southern Europe and the Iberian Peninsula, but as the ice melted the honeybee again extended its territory north and into the British isles, possibly aided by mankind's migration north which was happening at the same time. The native British black bee is closely related to those found in France and Spain.

Classification of the Honeybee

To understand the physiology and anatomy of Honeybees it is necessary to have some understanding of their classification (taxonomy) within the animal kingdom. Honeybees, species *Apis mellifera* belong to the genus *Apis*, which belongs to the family *Apidae*, which belongs to the sub-order *Aprocita*, which belongs to the order hymenoptera, which belongs to the class *Insecta*, which belongs to the phylum *Arthropoda* within the animal kingdom. I hope you took that in. You may need a second reading.

Like all members of the phylum arthropoda, which also includes mites, crabs, millipedes and spiders, honeybees have a body divided into segments. These are generally covered in a tough outer chitinous cuticle known as an exoskeleton, protecting the animal from predators, forming a container for the soft body tissue, forming a framework for the attachment of muscles and preventing loss of fluids. The segments may be fused together or connected by a narrow flexible membrane that allows movement of one segment relative to its neighbour.

The Honeybee

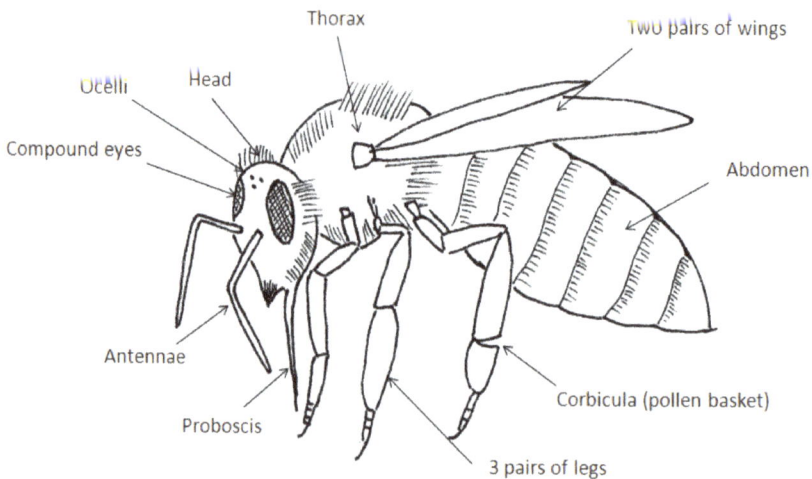

Like all members of the class of insects, which also includes butterflies and beetles, the body of the honey bee is divided into three sections, each section being made up of a number of segments, which is the same for all insects during their embryonic stage of development. The sections are the head (6 embryonic segments), the thorax (3 embryonic segments) and the abdomen (11 embryonic segments). In the adult, segments may fuse together and adapt to serve specific purposes. The head contains the mouth and the main sensory organs, such as the eyes and antenna. The success of the insect class in inhabiting so many niches around the world is largely due to the way the mouthparts can be adapted to meet the needs of different environments.

The thorax is made up of three segments, each carrying one pair of legs and, in some cases, wings. The thorax is therefore the section of the body responsible for locomotion. The abdomen contains the digestive organs, the stomach, the heart, the reproductive organs and the waste disposal organs.

Like all members of the order of hymenoptera, which also includes ants, the honeybee has two pairs of membranous wings, which are linked together by hooks, called hamuli. The males of the hymenoptera are haploid. This means that they have a single set of chromosomes. The females, like all mammals, most plants and other insects, are diploid, having pairs of chromosomes. So the queen and worker honeybees have 32 chromosomes, made up of 16 pairs, whereas a drone, the male, has a single set of 16 chromosomes. As a result each spermatozoa produced by a drone is a genetic clone of itself. This peculiarity results in evolutionary drivers that, it is thought, could explain the development of social insects, based upon a single queen.

Like all members of the sub-order aprocrita, which also contains wasps, honeybees have a petiole, a waist that appears to separate the thorax from the abdomen, though in fact the narrowing is at the second of the abdominal segments. And like all aprocrita the larvae of the honeybee are legless and rely upon the adults for nourishment.

Like all members of genus apis, which also contains bumblebees and solitary bees, the honeybee has adaptations on its rear legs to carry loads of pollen.

The anatomical specialisations of the honeybee, *Apis mellifera*, from other members of the genus *Apis* are generally minor and largely superficial in nature, such as size and colour. The factors that really do differentiate honeybees from other members of the *Apis* genus are behavioural. This is similar to the differentiation between Homo Sapiens and other members of the ape genus. Anatomically Homo Sapiens and apes are genetically and physically very similar – it is in the behavioural patterns that there are major differences, and these differences are linked to the size and complexity of the brain.

Species of the Honeybee

Within the species *Apis mellifera*, there are a number of subspecies, which have evolved in different geographical niches.

Apis mellifera mellifera – the black bee originating in Spain and Northern Europe and which is the native bee of the British Isles. These are large bees but with relatively short tongues. They winter well, are thrifty but tend to be nervous and overly defensive.

Apis mellifera ligustica – originating in Italy. These are smaller than the black bee but with longer tongues. They are distinguished with yellow bands on their abdomen. They are docile and have prolific queens. Because they are less thrifty, they tend not to over winter well in colder climates.

Apis mellifera carnica – a bee that originated in the Eastern Alps, but now widely found in Germany and central Europe. In size they are similar to the *ligustica* but are grey or brown in colour. They are a popular bee because of their docile temperament. They are prone to swarm readily.

Apis mellifera caucasica – originating from the Caucasian mountains. They are grey in colour, and reported to be slow to expand in the spring and are susceptible to Nosema.

Apis mellifera scutellata – originating from east Africa

Subspecies are able to cross breed with each other, and since the latter part of the 19th century honeybees have been moved about Europe and to America, as beekeepers have sought to breed the ideal bee. The Americas and Australasia did not have native honeybees, but bees were taken there by settlers in the 17th, 18th and 19th century.

It is not absolutely clear that subspecies can still be found in their areas of origin in their pure form. Certainly in the UK the majority of honeybees do not belong to the subspecies, *Apis mellifera mellifera*. Indeed some authorities even question whether there ever was a native breed of honeybee in the British Isles and suggest that honeybees arrived in the British Isles with the first human beings as they colonised the land as the ice retreated after the last ice age. There certainly is evidence that it was man that was responsible for taking honeybees to Ireland. But it is certainly true that *Apis mellifera mellifera*, the British black bee, was the predominant species in the British Isles for several thousand years prior to 1870 and as a result they became well adapted to cope with a harsher climate than the honeybees had to deal with in southern Europe. In the 1920's honeybees in the UK were decimated by a disease, known as the Isle of Wight disease, which entirely wiped out honeybees in some areas. The honeybee population was restored by importing honeybee queens from Europe, mainly *Apis mellifera ligustica*. This practice of importing foreign bees is still common in the UK and subject to much debate and some disapproval. Nevertheless, despite this history and current practice, there are small and isolated areas of England where the original black bee, *Apis mellifera mellifera* is still thought to thrive, such as on the Yorkshire Moors and in the Yorkshire Dales.

Physiology of the Honeybee

As with all animals the physiology of the honeybee needs to have a number of functions that include

Physical support

Respiration

Circulation

Ingestion and Digestion

Excretion

Locomotion

Senses

Nervous system

Reproduction

Glandular systems

These functions are common to all animals, including ourselves. It is interesting that, although the physiology of insects has the same functionality as the physiology of mammals, insects and mammals have evolved radically different methods to achieve the same ends.

Support

For animals that live on land, the soft tissue and organs need to be supported and protected and there needs to be a rigid structure against which muscles can be tensioned. Mammals have an internal skeleton. Insects have an exoskeleton that forms a rigid case around the body. The exoskeleton helps to conserve fluid within the body. The outer layers of the exoskeleton are not alive and are totally rigid. This raises a problem - to grow the insect must progress through a series of moults, and ultimately it restricts the size to which an insect, or indeed any arthropod, can grow.

Respiration

Mammals breathe in air through the mouth into the lungs, where oxygen passes through a thin membrane into the haemoglobin of the blood, which is then pumped by the heart to take oxygen to every extremity of the body. On the return journey to the lungs the blood absorbs carbon dioxide, a by-product of respiration. In the lungs the carbon dioxide passes back through the membranes into the air, which is then exhaled. By contrast insects have a network of tubes that carry air to every cell of the body, the whole body of the insect acting as a lung. These tubes are called trachea. The entrance to the trachea, where they pass through the exoskeleton, are known as spiracles and they incorporate valves that open and close in response to the insect's need for air. The trachea divide and subdivide into every extremity of the insects body, and are then known as tracheoles. The smallest tracheoles reach every living cell within the insect and oxygen passes through the cell walls into the cells, where it is used for respiration, providing the energy of life. At the same time, carbon dioxide is removed through the tracheoles. During flight, when there is an increased demand for oxygen, the abdomen can expand and contract to pump increased amounts of air through the trachea to the flight muscles. The abdomen can be seen pulsating when bees return to the hive after a foraging flight.

Circulation

Mammals have blood, which is pumped around the body by the heart. The blood is enclosed within a network of tubes, which are known as arteries, capillaries and veins. The capillaries divide and subdivide to reach every part of the body, and in this way nourishment, hormones and oxygen are transported to each living cell of the body. In addition waste products from respiration and cell renewal diffuse into the blood and are transported to the excretory system. In a similar way insects have blood, or haemolymph, which is pumped about the body by a heart located in the abdomen. As the haemolymph does not transport oxygen it does not contain haemoglobin and therefore is not red. Also, the blood is not contained in tubes for the complete circulation. The heart pumps the blood down the aorta towards the head, and then it washes through the body cavity, passing through the head, thorax and abdomen. Although the blood does not transport oxygen or carbon dioxide, it does transport nutrition, hormones and then removes the nitrogenous waste products of metabolism and cell renewal.

Ingestion and Digestion

Mammals ingest food through their mouths and it passes down the oesophagus to the stomach. The digestive process commences straight away, with enzymes breaking down the food into monosaccharides, simple fats and amino acids. The food passes through the small and large intestine where the digested food and fluid is absorbed into the gut lining and hence into the blood stream, and finally to the rectum and the undigested solid part of the food is eventually excreted though the anus. The physiology of honeybees has many similarities. The mouth parts are adapted to collect nectar from deep within flowers. The food passes down the oesophagus, which runs through the thorax, to the honey stomach, which is situated in the anterior part of the abdomen. The honey stomach is primarily a storage organ, storing nectar when it is being brought to the hive to be processed into honey. The honey stomach is also used to carry a store of honey when the colony swarms. It can contain up to 30mg of nectar, 25% of the weight of the honeybee and can expand to occupy half the space within the abdomen. The contents can be regurgitated. Small parts of the contents of the stomach can be allowed to pass, via a valve called the proventriculus, into the ventriculus, the main part of the gut, and then into the small intestine where the nutrition is absorbed through the wall into the haemolymph. Waste material from respiration and metabolism is added to the contents of the small intestine, before it leads into the rectum. The rectum is capable of expanding within the abdominal cavity so that the honeybee is able to retain waste material until the opportunity arises to void it outside the hive.

The Digestive System of the Honeybee

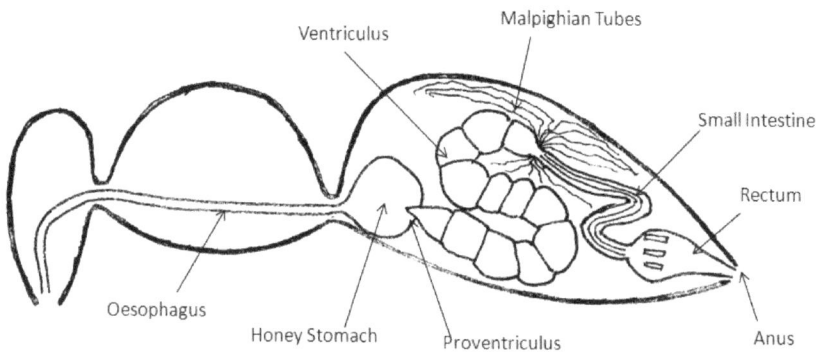

Excretion

Respiration and cell renewal is a continuous process within the living body, mammal or insect, and this produces waste material that must be expelled. In a mammal the waste materials, mainly nitrogenous compounds, are absorbed by the blood and then removed from the blood by filters in the kidneys. The kidneys also control the amount of water within the blood. The waste material produced by the kidneys is drained into the bladder and then periodically expelled by the body as urine, through a separate orifice. Similarly, for insects and honeybees the waste products of respiration and metabolism are absorbed by the haemolymph. These products are then removed from the haemolymph by an organ called the malpighian tubes, a mesh of fibrous tubes occupying the cavity between the various organs within the abdomen and connected to the small intestine at its junction with the ventriculus. The malpighian tubes filter the nitrogenous waste from the haemolymph and drain the concentrated waste products into the small intestine, and from there it is eventually excreted through the rectum and the anus.

Locomotion

The honeybee, like all insects, has six legs and two pairs of wings. All of these appendages are attached to the thorax. The space within the thorax is almost entirely filled with flight muscles, which, by rapidly distorting the shape of the thorax exoskeleton, cause the wings to beat, The wings beat at about 250 beats per second, and this is responsible for the characteristic buzzing sound of a honeybee, close to middle C. The bee is able to fly at about 24 km/hr. The aerodynamics of the honeybee flight is not totally understood, in that a satisfactory mathematical model has not been produced. It is probable that it is the eventual failure of the wings that results in the death of most worker honeybees, but during their lifetime they will each fly about 500 miles, the distance from London to Aberdeen. The thoracic muscles have another important function. By vibrating the thoracic muscles without moving the wings the honeybee can generate heat. This is one of the mechanisms that the honeybee uses to maintain a homeostasis in the brood nest and to generate heat at the centre of the winter cluster.

Senses

It is necessary for an animal to be aware of its surroundings, and to do this animals have senses. In the same way as mammals have a package of five main senses consisting of sight, hearing, smell, taste and touch, the honeybee has a package of senses – sight, smell, taste and touch, but not hearing. It is also thought that a honeybee can detect the levels of carbon dioxide and the earth's magnetic field.

The sight organs consist of two large compound eyes on the side of the head and three ocelli placed centrally, quite close together, at the top of the head. The ocelli are thought to be used as light detectors. The compound eyes consist of many thousands of small light detector cells, or ommatidia, arranged in an array. A drone has 70,000 of them, compared to a worker having 50,000. The picture of the world produced by these eyes is very grainy and not focused as we experience with our sight. However, the compound eyes are extremely sensitive to movement, to a far greater degree than the eyes of mammals. In addition, the ommatidia detect a different range of wavelengths to that detected by mammalian eyes. So whereas bees are unable to see red, they can detect ultra violet, which mammals cannot. The petals of some flowers have vivid patterns in ultra violet, invisible to us, but designed to guide the honeybee to the nectaries of the flower. Also, the eyes of the honeybee can detect the plane of polarisation of polarised light, an ability that is essential for navigation using the sun. The brain is located behind the eye and much of it is devoted to interpreting the images produced.

The antennae are the main organs of smell and touch, but the body of the honey bee is covered with setae, sense organs, which in various forms, like hairs in some cases, are used to taste, smell and touch. The senses of smell and taste aid the internal hive communications using pheromones, a function which is essential to the internal organisation of the honeybee colony.

The nervous system

The nervous system of a honeybee is highly developed. The main part is concentrated in the head as a brain, integrated with the main sense organs, the eyes and the antennae. But as with all insects there is a nerve cord running the length of the body with ganglions at intervals, which are concentrated groups of nerves that independently control the different parts of the body to which they are adjacent.

The Glands

The honeybee is a mini chemical plant, producing a large number of chemical substances. The glands can be categorised as

a)   Endocrine glands that produce hormones which circulate in the haemolymph and are part of the internal communication system in the body of the honeybee, in particular controlling development.

b)   Exocrine glands which produce substances that are excreted outside the body of the honeybee. Some of these substances are pheromones, which are used as a means of communication between the bees and play a significant role in controlling the behaviour of individual honeybees, and enable the colony to act as a unified whole. Examples of pheromones are the queen substance, the Nasanov (come hither) and sting (alarm) pheromones. In addition there are other exocrine glands that produce substances which are the raw materials of the honeybee economy. The hypopharyngeal and mandible glands situated in the head produce, amongst other things, food for the larvae and the wax glands, four pairs situated beneath the abdomen, produce wax.

The Castes of the Honeybee

It is commonly said that within the honeybee colony there are three castes, workers, drones and a queen. Strictly speaking drones are not a caste, just the male sex. The workers and the queen are castes. They have a similar genotype, both being female, but differ in their phenotype in that the workers are not sexually mature and have fully developed organs to produce wax, brood food and enzymes to process the honey.

Workers

The vast majority of the population of a honeybee colony is workers. In the early spring they number less than 10,000 but in June and July the number of workers will peak at between 40,000 and 60,000. The workers are responsible for all tasks within the colony, except reproduction. The first three weeks or so of their adult life are spent within the hive, initially responsible for cleaning, then feeding the brood, processing honey and building comb. In the second part of their life their duties lie outside the hive, guarding the colony and foraging, bringing back the nectar, pollen, propolis and water that the colony requires.

The organs within the body of the honeybee continue to develop after the worker emerges as an adult and these developments are mirrored in the changing roles that the worker carries out within the colony. The hypopharyngeal gland found above the mouth parts first produces brood food for the developing larvae. Later, when the worker becomes a forager the same gland produces invertase, an enzyme which is used to break down the disaccharides in nectar into monosaccharides, part of the process of converting nectar into honey. The wax glands become active after the worker is about ten days old and normally cease to produce wax after another week. The worker emerges with no venom in its venom sac but over the next three weeks the venom is produced and the sac is full after this period. It is at this point in the worker's life that she may act as a guard bee.

In the summer their adult life is no more than 40 days, dying when their wings disintegrate. It is thought that a worker flies about 500 miles in its lifetime. In the winter workers can live for six or seven months, but at a much lower level of activity.

Drones

The male honeybees are known as drones. They are produced in varying numbers during the summer months, from April through to August, their numbers peaking at less than 15% of the colony population. At the end of the summer they are expelled from the hive and perish. They have a single role and that is to mate with a virgin queen, and the vast majority fail to carry out this task. Their body is designed to give them the optimum chance of being able to fulfil this mating role. As mating takes place in the air 10 to 30 metres above the ground and well away from the hives in what are called drone congregation areas, they need to be powerful flyers, requiring the broad thorax containing the flying muscles. Enhanced sense organs, in particular the eyes and antennae, enable them to detect and then catch a virgin queen. The act of mating results in the death of the drone.

Queen

For most of the time there is a single queen in a colony. Genetically she is identical to a worker, but having been fed richer and greater quantities of food as a larva she is larger and has reached full sexual maturity. Her abdomen, which contains the ovaries, is about 30% longer than that of a worker so that her wings extend just two thirds of the way down the length of her abdomen. She has a sting but it is only used when fighting with another queen. Her major role is to lay eggs. At the beginning of her life she will mate with several drones over a period of a few days. The sperm she receives at that time, up to seven million, is stored in a spermatheca, and released in small numbers to fertilise the eggs that she produces over the next few years. At the peak of the summer she is capable of laying between 1500 and 2000 eggs each day. A queen can live up to four years, but usually her life is two to three years. The queen has a secondary role. By producing pheromones she promotes colony cohesion. I said that for most of the time there is a single queen. The main exception to this is during periods of swarming, which will be described elsewhere.

As the queen lays the eggs she appears to be able to control whether or not to release a small number of sperm, the vehicle carrying the male gamete, from her spermatheca, to fertilise the egg as it passes through the oviduct. If the egg is to be laid in a worker cell or a queen cup then sperms are released, but if it is to be laid in a drone cell then the sperm is withheld. In simplistic terms, the sex of the honeybee is determined by whether it is haploid or diploid. Being haploid means that it has a single set of chromosomes (16 in number) and this results in a male. Being diploid means that it has two sets of chromosomes (32 in total) and this results in a female. Before laying her egg the queen can be seen assessing the diameter of the cell with her front legs. Drone cells are larger than the worker cells.

Development of Brood

The time taken for the egg to develop into an adult bee, passing through a number of critical stages, is essential knowledge for the beekeeper. Though a little complicated when first met, the facts involved soon become second nature. Without an awareness of these timescales it is not possible to have an understanding of swarm control procedures or to master queen rearing methods.

All eggs take approximately 3 days to hatch. The queen lays the eggs so that they stand parallel to the axis of the cell and then they slowly lean over. The larva is essentially a gut with a simple mouth part at one end. It eats and it grows at an impressive rate. The larvae are fed by workers who are themselves are just a week or so old. The brood food is manufactured in glands in the head of the bee. It is protein rich and to obtain the protein the workers eat pollen. It follows that an expanding colony must have access to ample sources of protein. During development the juvenile honeybee moults six times, five of these occurring during the larval stage. The larval stage lasts 5.5 days for the queen, 6 days for the worker and 7 days for a drone. During those few days the weight of the larva increases by 900, 1700 and 2300 times the weight of the egg for the worker, queen and drone respectively.

The third stage is the pupal stage. At the end of the larval stage the larva moves to lie length ways along the cell, head outermost and spins a cocoon about itself. Rather distastefully at this point it releases the built up quantity of faeces from its gut and this is incorporated into the cocoon. The workers cap the cell with wax and in the next few days a remarkable change occurs. The larva metamorphoses from a simple grub into an adult bee, most of the body of larva being broken down and then rebuilt. For a queen the process takes 7.5 days, for a worker 12 days and for a drone 14 days.

Time Line for Honeybee Development

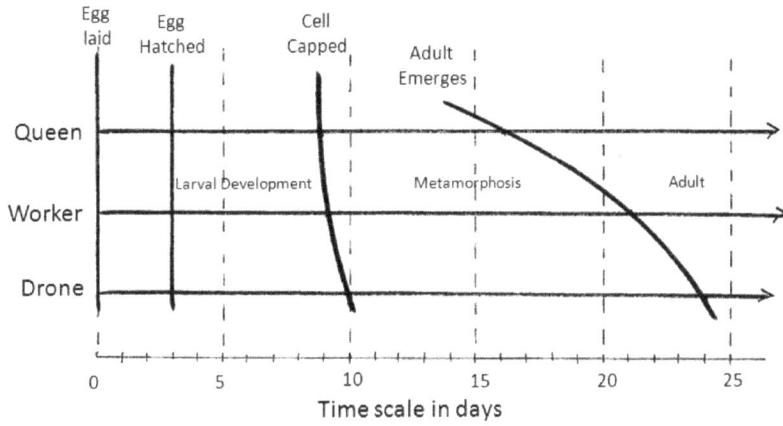

In the following chapter I will look at the biology of the honeybee superorganism.

Chapter 3
The honeybee as a superorganism

Species and Subspecies

A species can be defined as a group of animals or plants that is able to successfully interbreed, producing viable and fertile offspring. There is considerable variation within a species and geographic and other forms of isolation can result in several subspecies, each with distinct characteristics but still having individuals that are able to interbreed with individuals of other subspecies within the species. Even within the subspecies there is still a great deal of variation. This variation is the raw material of natural selection and allows species to adapt to their environment as their environments change.

Sexual Reproduction

In the following paragraphs I argue that honeybees have evolved to become one of the most advanced social animals on earth, and by virtue of this able to adapt to a wide range of environments. If this conclusion suffices then please feel free to skip quickly through the next five paragraphs. But if we get down to details...

Chromosomes are thread like structures observed during the reproduction of cells and are composed of sequences of DNA in the form of a double helix. These combinations of DNA specify genes, which define or control the physical characteristics of the organism. Every genetically defined characteristic is controlled by a gene or set of genes, and these have fixed loci or positions on the chromosomes. But for each characteristic there can be many alternative forms of the genes. These different forms are called alleles. The example with which we are probably most familiar is the colour of human eyes. Alleles exist for blue eyes, green eyes, brown eyes, hazel eyes etc. For all mammals and most insects there are two sets of chromosomes (diploid) and therefore each individual has two alleles for each characteristic, only one of which normally is triggered. This allele is said to be dominant and defines the characteristic in the individual's phenotype. The phenotype is the physical manifestation of the genotype. In most living organisms there are hundreds of thousands of genes defining the phenotype of the species, and each of these genes will have countless differing alleles. For diploid species, both animal and plant, which use sexual reproduction, male and female produce gametes which contain one of each pair of chromosomes (haploid). The male gamete is called a sperm and the female gamete is known as the egg or ovum. The process by which these are formed, meiosis, involves a recombination of the genes within the chromosomes so that the probability of there being exact duplicates of either sperm or ovum is extremely small. And so when the male and female gametes combine, during sexual reproduction, to form a zygote (once again diploid) that will develop into a new individual, that individual is genetically unique. Subsequently that individual may be cloned to form genetically identical individuals such as when identical twins are formed. The sperm of haploid organisms such as the honeybee drone, are genetic clones of the parent.

Mutation and Evolution

At a much slower rate, random mutation within the genome is taking place. Each mutation effectively will form a new allele of a gene. The vast majority of mutations do not occur in cells that are involved in reproduction or result in phenotypes that are not viable. Either way these mutations will never be transmitted to offspring and will never become a part of the genetic makeup of the species. But occasionally a mutation will occur producing an allele that results in a phenotype that has an advantage over its competitors and is well adapted within its niche, and so that allele, belonging to successful individuals, will become widespread. Sexual reproduction keeps producing and testing the possible variants of the genotype of the species and from this vast palette nature selects the individuals which are best able to survive in each niche. There is no final perfect outcome as the environment is never constant.

The number of niches or environments provided by the earth is vast and forever changing. For instance in the last 10,000 years (a blink of the eye in geological terms) the lowland landscape of the British Isles has evolved from tundra to birch forest to oak forest to grassland, monoculture agriculture and urban gardens that we see today. Natural disasters and the effects of competing species can cause even faster changes. We are aware of the accelerating rate of global warming, most likely due to the excesses of mankind, but the climate has never been totally stable.

If a species fails to adapt sufficiently as its environment changes then it becomes extinct. At the same time new species emerge from subspecies, when, due to environmental or geographic isolation, they diverge genetically to a sufficient extent from their original species so that they are no longer able to interbreed. Species have differing methods of adapting to the changes of environment. At one end of the spectrum, species reproduce to excess, either by each individual producing vast numbers of offspring or by reducing the time for each life cycle. In this scenario it only requires a small proportion of the offspring to survive to a point where they themselves can propagate for the species to be successful. The production of vast numbers of offspring increases the probability of a genotype being produced that is well adapted to its niche. It also increases the chance of genetic mutations.

Social animals

Towards the other end of the spectrum we have human beings. Their life cycle is 20 to 30 years and each female will only produce a small number of offspring in her lifetime, less than two on average in modern developed countries, and historically less than ten. As a result, mutation of the genotype is very slow, much slower than the rate of environmental change. But human beings have other means of adaptation. Through their social and individual intelligence, they are able to alter the immediate environment that surrounds them by making clothes, building homes, producing food through agriculture and manufacturing tools. As a result. they can adapt and survive in a wide range of natural environments, and now inhabit much of the earth's land surface.

Similarly, honeybees are able to adapt to their natural environment in which they live by producing their own environment within the hive, a manufactured environment constructed out of wax comb and regulated by the honeybees' ability to ventilate and heat the hive, maintaining a controlled atmosphere within the hive, which, to a significant extent, is independent of the external environment outside the hive. And so, like human beings they are able to survive and prosper over much of the earth's land surface. In both cases the ability to adapt to different environments comes from cooperation and social interaction, rather than genetic selection.

There is a significant cost to the honeybee (and human being) of maintaining this ability to flexibly adapt to any environment where it finds itself. As this ability to survive is based on social organisation the individual cannot survive alone, and so propagation requires large numbers of individuals to be involved. And because sophisticated communication is required each individual requires a highly developed nervous system and brain, and this requires a significantly greater investment in raising the young. As a result, both human beings and honeybees have long life cycles compared to other members of their class in the animal kingdom, and each life cycle produces just a small number of offspring.

The Superorganism

Just to be clear, the life cycle of the honeybee should not be thought of in terms of the individual bee but in terms of the colony, the superorganism. Though the queen can lay between 1500 and 2000 eggs per day during the height of the summer, the vast majority of these develop into workers, which, with a minor exception of drone laying workers, are genetic dead ends. From the colony's point of view, the production of workers is simply a renewal of the working parts of the superorganism, similar to the continual renewal of cells in the body of a mammal. In her lifetime a queen will probably produce less than 10 viable queens which go on to lead a colony. Much less than this in most cases.

A superorganism is an entity made up of many thousands of individual organisms, but for the purposes of understanding its biology it needs to be considered as an indivisible, integrated whole. In many ways the superorganism has the properties of a mammal. It reproduces at a slow rate, provides its offspring with nourishment, has within its body a sophisticated nervous or communication system and maintains a body temperature of about 35°C.

The ability to swarm, to build wax comb, to store honey, to store pollen, to regulate the temperature of its brood nest, to defend itself are all properties of the honeybee superorganism rather than of the individual honeybee. The key to the ability to exist as a superorganism is communication between individuals within the superorganism. The mechanism of communication for human beings is primarily speech, though pheromones and visual signals may play a significant role, more than we sometimes recognise. For the most part, however, the communication skills are learnt. By comparison the mechanisms of communication between individual honeybees include dance, pheromones and trophallaxis (the sharing of food). These behavioural characteristics are innate, rather than learnt, but like the speech of human beings they require a greater development of the brain and nervous system than other species in the same class of animals. For many years, scientists have puzzled over why the honeybees invest so much energy in maintaining the brood temperature at about 35°C, very close to the human beings body temperature of 36.9°C. It is now thought that this temperature is required so that the brain of the honeybee will develop sufficiently to be able to successfully communicate.

The ability to communicate and social cooperation have evolved together, and social cooperation leads to the ability to adapt to one's environment. Human beings, without the ability to cooperate with their fellows, would never have migrated to the more temperate parts of the world. Without the social cooperation it would not be possible to build shelters, store food and make clothing, skills that are required in order to survive in temperate regions. In the same way honeybees would never be able to survive the winters of the temperate regions without the cooperative behaviour that enables them to build comb, store food and maintain a hive temperature. As we regard honeybees as one of the most sophisticated and adaptable of insects, I would hope that they regard us in a similar positive light!

For many centuries it was believed that a colony of honeybees was led by a king bee and then, when it was observed that the 'king' bee was laying eggs it was then acknowledged that it was a female who was in charge, that is the queen. But in fact, there is no leader, no directing intelligence. Each individual worker makes decisions as to its behaviour and actions in accordance with a set of genetically inherited rules or algorithms and in response to stimuli from its environment and the other workers. This incredibly complicated mesh of stimuli and feedback results in the behaviour patterns that we observe. We are far from fully understanding them. The queen does play a unique role. She lays the eggs and produces pheromones, in particular the queen substance, which is passed throughout the colony, but she is not the source of a governing intelligence.

I will examine in detail some of the features of the honeybee superorganism, the comb, the ability to control the temperature of the hive, defence strategy and foraging strategies. I'll also be looking at swarming, the reproduction of the superorganism, but that is in a separate chapter.

The Annual Cycle

But first I will describe the annual cycle. In the middle of the winter the colony forms a cluster and reduces its temperature so that it can continue to survive while minimising the expenditure of energy. However, after the winter solstice the temperature of the central part of the cluster is raised to 35°C and in this restricted area the queen will start to lay eggs. For a couple of months, the energy and protein required to nourish the new brood will come from the resources stored during the previous summer and autumn. As the temperature outside the hive increases and early spring flowers come into blossom, some of the bees will start foraging and the area of brood increases. Despite the increasing number of new bees being produced the total population of adult bees will continue to decrease until the beginning or middle of April, reaching a minimum of about 10,000. From then on there is a steady and accelerating increase in the population.

Towards the end of April, the colony will start to produce drones and then from the middle of May through to the end of June is the time of year when the colony is most likely to produce swarms, though swarms can be produced any time from April through to August. The population of the colony will continue to increase through May and June and peak during July, typically with 50,000 to 60,000 bees. During the summer months the bees will, if the weather conditions are reasonably favourable, produce a surplus of honey. When the peak of the honey flow occurs very much depends upon the local flora. In my neck of the woods in Yorkshire it was from the middle of July through to the middle of August, exploiting first the blackberry and then the Himalayan Balsam and Rosebay Willow herb.

During August, September and October the queen gradually reduces the rate at which she lays eggs and by the end of November she may have ceased to lay altogether. Once the temperature drops to 18°C the bees form a cluster and the cluster contracts upon itself as the temperature falls further. From the middle of October through to the end of March the colony depends upon the honey stored during the previous summer.

The annual variation of the population is illustrated on the graph. April can be seen to be a particularly stressful period when the amount of brood equals or even exceeds the number of adult bees. A failure to be able to access sufficient forage during this period can lead to disease, a failure to build up adequately which will affect the colonies performance throughout the subsequent year and may even result in a complete collapse through starvation.

Annual Variation in population of adults and brood

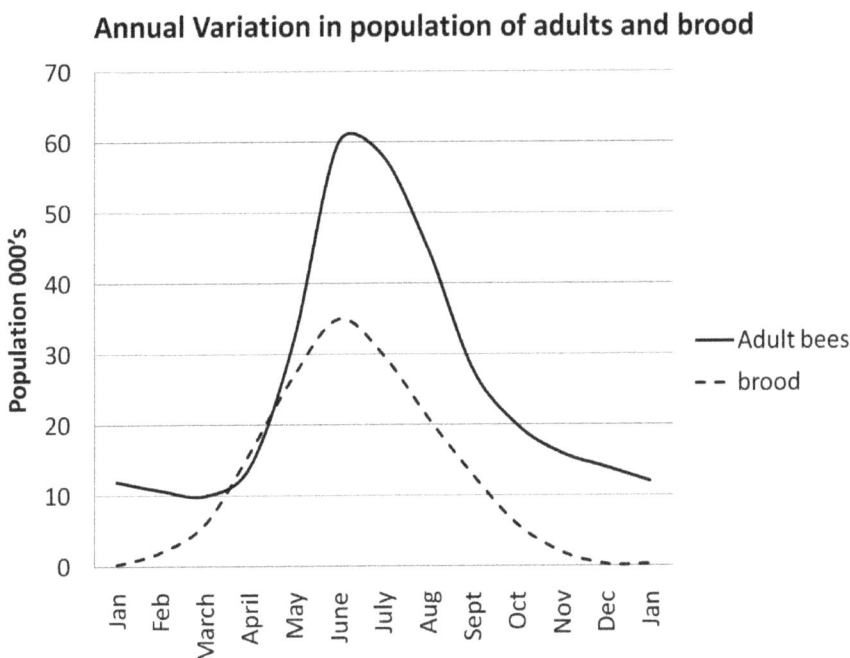

The Comb

The comb is an integral part of the honeybee superorganism. The comb is built from wax produced in the four pairs of wax glands situated beneath the abdomen of each worker. The scales of pure white wax are transferred by the legs (the rear legs have an adaptation to handle the wax scales) to the mandibles where they are warmed and mixed with saliva until the wax is sufficiently malleable to mould and then is added to the wax comb under construction. The wax glands convert the carbohydrates originating from ingested honey to produce the wax. About 8kg of honey is required to produce 1kg of wax. That 1kg of wax is sufficient to produce 30,000 cells.

When a feral swarm occupies a hollow cavity, the worker bees will start the construction of the comb by attaching a ridge of wax, strengthened by propolis, to the roof and then as more and more wax is added the comb is gradually pulled out into hexagonal cells. Once the comb is over 2cm deep it starts to conform to the rigid specification of the honeybee comb. The combs are not necessarily built in truly plane lamina, but they are precisely vertical and there is a fixed gap between adjacent combs. On both sides of the comb there is a tessellation of hexagonal cells, the axis of each cell at 13° to the horizontal, sloping upwards from the septum. The walls of the cells are precisely built with a thickness of 0.07mm. To maximise the strength of the comb, the cells on one side of the comb are offset in relation to the other side so that the vertices on one side correspond to the centre of the cell on the other. The rims of the cells are thickened and apparently reinforced with propolis. The wax comb of the honeybee is indeed one of the marvels of the natural world.

The majority of the cells are worker cells and these are about 5.2 – 5.4mm in diameter. Drone cells are somewhat larger, 6.2 – 6.1mm in diameter. When the bees are given wax foundation they will accept and build cells upon the hexagonal tessellated pattern which is pressed into the wax on both sides of the foundation. Later they may convert some areas of worker comb to drone cells as required. Most of the comb is reused repeatedly, up to seven or eight times each year for cells at the centre of the brood nest. Before reuse the cells are polished and coated with propolis and the remnants of the cocoon, incorporating it into the cell structure, so that, in time the internal dimensions of the cells decrease as the walls become thicker. As a result, when old brood comb is melted the wax runs off to leave a dark brown skeleton of the comb, formed in propolis and other material.

There is a common paradigm in the way that the bees utilise the comb. In the middle, towards the bottom, is the spherical brood nest. In feral colonies that are still growing the nest is always in the newest comb towards the bottom. Towards the periphery of the brood nest, depending upon the time of year, there may be areas of drone brood. It is known that drone brood can develop at a slightly lower temperature. I could speculate that this is because drones are not required to develop the behavioural complexity of the workers. (For female readers I request that they don't take the analogy I've been using between human society and the honeybee colony too far). Outside the brood area is an area of pollen storage and beyond that and above is the honey comb.

Homeostasis

The wax comb gives the colony shelter and protection. In the summer it is not uncommon to find honeybee nests out in the open, possibly in a tangle of branches within a thorn tree. This is the result of a swarm which has failed to find a cavity in which to make a permanent home. During the summer months these colonies can grow and prosper. Of course, once winter sets in they are unlikely to survive. But even to prosper in the summer months in this country where the average summer temperature is well below 20°C, they need to permanently maintain a brood temperature of 35°C in the brood nest, over 15°C above the average of the external environment. This is a clear illustration of how the comb provides a high degree of protection and insulation.

Most colonies live inside hives or natural cavities and in there the comb is sufficient to enable the colony to survive throughout the coldest months. The wax comb traps air and that makes it into an excellent insulator. A comb full of honey is even better. The vertical laminar columns of air between the combs can be managed by the bees in various ways. The bees can either use their bodies as a cluster to prevent the flow of air up through the columns, they can allow the air to flow naturally by convection up the chimneys of air between the combs, or they can actively encourage the air flow by fanning their wings. The ventilation and temperature of the brood nest is under the control of the bees. The colony has two other strategies at its disposal, to deal with the more extreme conditions. When it is cold, the workers are able to generate heat by vibrating their flying muscles in the thorax, an activity which is fuelled by honey. And then in the extreme heat of the summer the bees bring water into the hive, spread it over the combs in a thin film, and allow evaporation to lower the temperature so that it is within the required limits. Temperatures that are too high are just as fatal to brood as temperatures that are too low

Whenever there is brood, the bees try to maintain a homeostasis, a stable state as regards both the temperature and humidity. Honeybees are able to achieve this to a greater degree than any other insect, and this ability is the single most significant factor which enables the honeybees to prosper in many different climates. During the spring and summer months the brood nest temperatures can be maintained at about 35°C with a variation of less than 1°C. At the edges of the brood nest, where drone brood is raised, the temperature is allowed to drop to about 30°C. The temperature in the supers is not so tightly controlled.

When the outside temperature falls below 18°C the bees start to cluster, and so it should be understood that clustering behaviour is not confined to the winter but also to summer nights. The cluster forms a sphere encompassing and centred on the brood nest. The outer layer forms a barrier to the movement of air. As the temperature falls the cluster contracts, but most of the time there is a core at the centre of the cluster which is less densely populated and where the queen can lay. During the depth of the winter the colony allows the core temperature to fall below 30°C and for a period the production of brood is suspended. This period is variable and to some extent depends upon the severity of the winter. Some beekeepers reported that during mild winters brood rearing continued right throughout the winter. Once the queen ceases to lay the core temperature can drop to 20°C. This is the lowest temperature which will guarantee that the bees at the periphery are kept above 8°C. This is necessary, as below this temperature, the outer bees can fall from the cluster and perish. By reducing the temperature to this minimum it minimises the use of the honey stores.

Communication

So we can see that the comb is integral to the maintenance of homeostasis. The ability to communicate is another primary requisite of the honeybee superorganism and the comb is integral to this function. There are two main methods of communication within the superorganism, dancing and pheromones. The comb provides the dance floor for these dances. The best known of these dances is the waggle dance.

Unlike bumblebees and other insects that collect nectar for their immediate needs, the honeybees' life cycle requires them to collect enough nectar to be converted into honey, a food reserve which, in sufficient quantities, will allow the colony to survive the winter, a period equal to half the year in temperate regions when there is little or no nectar available. Because of this enormous requirement for nectar, honeybees have evolved sophisticated strategies for efficient foraging. When a source of nectar is discovered by a worker, it is essential that it is exploited fully by the colony as a whole, and when a source of nectar becomes exhausted it is essential that the foraging resources of the colony are quickly redirected to fresh sources. This all sounds like the language that might be used to describe the activities of a multinational company, but the flexibility that honeybees display in exploiting their resources is far more impressive than any multinational. A colony of honeybees is capable of redirecting its foraging resources two or three times in a single day. The potential number of bees in a colony available for foraging is half of the adult bee population, that is, in the middle of the summer, about 25,000 bees in a strong colony. Experiments have shown that honeybees regularly can fly 6 km (3.75 miles) from the hive in search of sources of nectar, and so it follows that their potential area of operation is over 100 km² ($A = 3.142 \times 6^2$). Having said that, the majority of foraging takes place within a radius of 2km (1.25 miles) from the hive. This still represents an area of operation of 12.5 km². When this is considered it can be seen that a worker honeybee that discovers a good resource of nectar, must have some method of communicating the position and the value of that resource to her sisters back in the hive. Even without knowing the nature of the communication, we can deduce that this communication must exist. All beekeepers are aware that if they leave an exposed super full of honey a couple of hundred metres from the apiary, within an hour there are thousands of bees taking the honey and by the end of the day the combs will be empty. This couldn't happen by chance.

It was Von Frisch in Germany in the 1940's who discovered and interpreted how this communication works – the wonderful waggle dance. There are three items of data that need to be transferred from the returning forager to the candidates that must be recruited to exploit the resource.

a) The direction of the source from the hive

b) The distance of the source from the hive

c) The quality of the source

The Round Dance

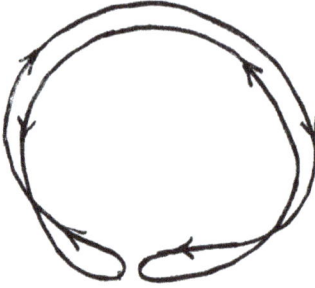

There are a number of forms of the dance. To communicate that a source of nectar exists less than 15m from the hive the round dance is performed. This gives no indication of distance or direction. When a source of nectar suddenly appears, such as if there is a spill of sugar syrup during feeding, this dance can result in a frenzy of activity in the apiary as each individual bee tries to locate the source. For this reason it is best to feed in the evening.

Beyond 15m up to 100m there are intermediate forms of the dance. The waggle dance in its full form is used when the nectar source is more than 100m from the hive. The dance consists of the returning bee running in a pattern consisting of two loops, alternatively clockwise and then anticlockwise, forming a figure of eight. Between the two loops there is a straight run. It is this straight run which encodes the data being transmitted.

The angle that the straight run makes with the vertical encodes the direction in which the source of nectar is from the hive relative to the sun. "Amazing!" you say. But there is no alternative. The bees haven't got a compass or GPS system, so the sun is the only thing that the bees could use as a basis of navigation. After some thought you then point out that the bees still seem to forage effectively when it is cloudy. And yes, they can do this and this is because their eyes can detect the pattern of polarised light that the sunlight produces as it passes through the atmosphere and from this pattern deduce the position of the sun in the sky. The pattern of polarised light is present even if the sun is obscured by clouds.

As the dancing bees make the straight part of the run the bee waggles its abdomen from side to side, and at same time buzzes. The attendant bees can deduce the distance from a number of factors,

a) The number of complete waggles performed on the straight run. The more waggles completed per circuit, the greater the distance. So for instance, 10 waggles per run would indicate a distance of

0.5km, 28 waggles per run would indicate approximately 2km. The relationship is not linear, and it appears that in fact it is not strictly speaking the distance that is indicated but the effort required to fly there. So the distance indicated for a given source will vary depending upon whether there is a head wind or not.

b)   The time for a circuit. The shorter the time for a circuit then the shorter the distance.

c)   The duration of buzzing when the bee is doing the straight run. The longer the duration of the buzz the longer the distance to the nectar source.

The Waggle Dance

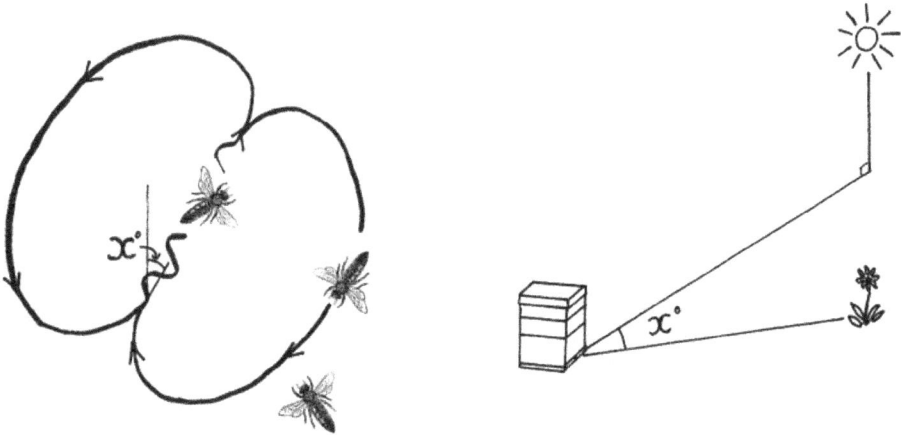

The dances are usually performed on what seems to be a designated area on the comb, which is usually fairly close to the entrance. The dances can sometimes be observed when examining a hive. It needs to be remembered that the dance is performed in absolute darkness so the attendant bees are not seeing the performance. It is more likely that they are using the properties of wax comb which allows it to transmit the vibrations of the dance to the attendants.

The enthusiasm of the waggle gives an indication of the quality of the source, but more importantly the dancing bee occasionally stops dancing and offers the spectator bees samples of the nectar, and not only does this show to the candidates the quality but it allows them to learn the associated odours of the nectar which will help them in finding the nectar source should they decide to be recruited to exploit the resource.

Defence

Except for the queen, the honeybees' top defence priority is the colony, not the individual bee. As the workers mature, they take on different roles in the colony. Workers become guard bees when they are about twenty days old, and this corresponds to the age when the amount of venom in the venom sac is a maximum. The amount of venom gradually reduces as they take on the role as foragers. The sting is an adaptation of the ovipositor. When attacking, the worker exposes the sting and begins to release sting pheromones. She arches her abdomen and a system of levers drive the lance of the sting into the flesh of the intruder and barbs hold the sting in place. When the worker tries to disengage, the barbs prevent the sting being removed and the complete sting organ is left behind and the bee, mortally wounded, dies. While still attached to the victim the sting continues to pump venom through the lancets into the victim. At the same time the scent glands attached to the sting are pouring pheromones into the air and attracting more guard bees to the sting site. If you get stung and you can smell the pheromone, isopentyl acetate, then it is time to retreat. It is suggested that the fact that the worker honeybee dies after stinging is an accident of evolution. I think not. It is the most effective way of defending the colony. The venom is used fully and reinforcements are quickly attracted to drive home the attack. The loss of several workers out of 50,000 is insignificant. By contrast the queen only uses her sting to kill rival queens and the barbs on her sting are not large enough to prevent withdrawal.

Chapter 4

Swarming

The propagation of the honeybee superorganism, the colony, is achieved through swarming. In simplistic terms swarming occurs when the colony produces new queens and the colony divides, part of it staying in the original hive and part leaving to establish a colony in a new location. Swarming requires an immense investment on the part of the honeybee colony and therefore only occurs when the conditions are appropriate. Without swarming honeybees would have no mechanism to expand into new territories and replace colonies that have been subjected to predation or natural disasters such as fire. So swarming is a process that is absolutely essential to the survival of the honeybee species.

Despite this beekeepers tend to regard swarming as an inconvenience, something to be controlled or even prevented. It is true that a colony that has swarmed will produce less honey and it is also true that swarms can be regarded as a nuisance by the public and cause alarm. Like most beekeepers I practice swarm control because I want to keep colonies that will produce honey, but I don't get overly concerned if I do occasionally lose the odd one. It may be argued that losing a swarm into the wild does have a benefit to the long term wellbeing of the honeybee, if not to the individual beekeeper who now has a depleted colony. When the varroa mite arrived in this country – it was first detected in 1992 - there was a significant reduction in the number of feral colonies. Whereas beekeepers are able to put into place husbandry practices that enabled their managed honeybees to coexist with varroa, the majority of feral colonies appeared to have inadequate defences against varroa and most eventually died.

Varroa is a mite that for millennia coexisted with the South East Asian honeybee *Apis cerana*. One long term hope for dealing with the varroa problem is that *Apis mellifera* will adapt through natural selection so that it too can coexist with varroa. In circumstances where beekeepers are continually treating their bees with medications to control varroa it is difficult for natural selection to work. A small number of beekeepers and researchers are engaged in breeding programs to produce bees which can coexist with varroa. But in the wild, amongst the feral colonies, natural selection will slowly, over a long period, and inevitably lead to the evolution of a variety of honeybees that can coexist with varroa. In addition feral colonies contain a pool of genes that cross breed with the kept honeybees, producing a vigour that would otherwise be missing. Whereas beekeepers breed for good temperament and good honey production, in the wild, natural selection favours the bees that simply survive by being able to adapt to the local climatic conditions, being prudent with stores, and developing immunity to disease. So if we lose a few swarms into the wild it's no bad thing. We are ensuring that the population of feral bees is maintained and we are adding to the feral gene pool.

In a given year, it is by no means inevitable that a colony will swarm or attempt to swarm. Most swarming occurs during May, June and July, but the swarming season can extend from April through to August. As mentioned before, the group of pheromones, known as the queen substance, is produced by the queen. Queen substance is distributed throughout the colony by the workers, through grooming, physical contact and trophallaxis, which is the sharing of food. It is thought that the presence of these pheromones in sufficient concentrations inhibit the workers from producing queen cells. Should the amount of queen substance become insufficient the workers will produce queen cups, the first sign of swarming. This lack of queen substance can occur if the colony grows too large, becomes congested or if the queen, through age, is unable to produce sufficient queen substance. Therefore, the risk of swarming can be reduced by

a) Using queens less than two years old
a) Ensuring that there is always sufficient space within the hive for the bees.

Unlike worker and drone cells, swarm queen cells are built to point vertically downwards. When sealed they are about 25mm long and pitted in the same way as a peanut. On wild comb the swarm queen cells are built hanging from the lower catenary curve of the comb. In hives with moveable frames, this curved lower edge does not exist and so the queen cells are built in holes that the bees have nibbled at the edge of the comb or are built at the bottom or on the side of the comb so that they are on the face of the comb and sometimes overlapping the bars of the wooden frame. Typically there can be between ten and twenty swarm queen cells spread across several brood frames. Swarm cells can be preceded by queen cups that look like the cup of an acorn, about 6mm across. The presence of queen cups is not in itself an unambiguous indication that the colony is about to swarm unless it can be seen to contain an egg or larvae. The workers either induce the queen to lay in them or move a newly laid egg into them.

Three days after the egg is laid it hatches. The workers then feed the larva with a rich diet known as royal jelly, feeding them with as much as they can ingest. As the larva grows the walls of the queen cell are extended downwards until they are about 20 - 25mm in length. After between five and six days the larva is fully grown. The larva positions itself length ways in the cell, head downwards, the cell is sealed and in the next eight days the larva undergoes a final moult and metamorphoses into an adult, virgin queen.

Once the workers recognise that there are queen cells developing with a viable larva, they reduce the amount that is fed to the old queen. As a result of this and other disruptive behaviour on the part of the workers, she stops laying and loses weight. Meanwhile scout bees start to look for a suitable site for the new colony that will eventually emerge. The new site must be a cavity, about 40 litres in capacity. Ideally it needs to have a small entrance, which is several metres above the ground and which faces south. The cavity needs to be weatherproof. In the natural state hollow trees are ideal, but small rock cavities, unused chimneys, cavities behind fascia boards and cavity walls are often utilised. Of course, empty beehives also make an ideal location, especially if they are placed on a flat roof. In nature the purpose of the swarm is to spread the bees into new areas and therefore the scouts will generally seek a new home several hundred metres from the site of the original hive. Beekeepers often leave bait hives in their apiaries to collect swarms. In general, this is not going to help collect swarms from your own apiary, but it is possible that you will receive swarms from neighbouring apiaries. If you have an isolated apiary it is better policy to place the bait hive a 100 metres or more from the apiary rather than in the apiary itself.

Provided the weather is suitable a swarm will leave during the day before the queen cells are sealed. This is called the prime swarm. Later the colony may issue afterswarms or casts, but it is the prime swarm that is most valuable. It contains large numbers of workers and these workers are well provisioned. The prime swarm has a good probability of getting established and surviving the following winter. The prime swarm consists of about 60% of the workers of the parent colony, a cross section of the different age groups, but without the very youngest or oldest. The swarm is accompanied by the old queen, now slimmed down sufficiently so that she can fly. Some drones may join the swarm. Prior to departing, the workers fill their honey stomachs with about 30mg of honey, and this store of provisions will provide for the swarm while it is being established in a new home. The first movement of the swarm is quite a short distance, often less than 20 metres. If the bees have read the books they will form a cluster, hanging from a branch about two metres above the ground. The cluster forms an oval shape, like a rugby ball, the length of which depends upon the size of the swarm and therefore the weight of the bees. In a long established apiary swarms will often go to the same place time after time, attracted presumably by the pheromones left by previous swarms. But in reality swarms end up in all sorts of less convenient places. I've collected them from fence posts, the sides of planters, on garden furniture, walls of houses, deep inside thorn bushes and on factory walls. Every year there are newspaper stories of swarms that have appeared in even more inconvenient places, such as on traffic signs and within vehicles.

A swarm in a willow tree at the rear of the apiary

A swarm that didn't read the book

Having swarmed, the process of scouting for a new home continues apace. This can take a few hours or several days, and in a few sad cases the bees totally fail to find anywhere suitable and start to build combs at the site of the cluster, where in all likelihood they are doomed to die during the winter. Returning scouts can communicate to the swarm that they have found a suitable place by performing a waggle dance on the surface of the swarm, and in so doing recruit additional scouts to review their proposed site. Initially several possible sites may be under consideration, but eventually a consensus is reached. Even without visiting the apiary or seeing a swarm, the beekeeper will detect that swarming is underway by the unusual behaviour of the scout bees, flying around the eaves of the house, investigating corners of the shed etc.

Having reached a consensus the swarm becomes airborne and flies in a spectacular dark cloud of insects to its new home. The scouts buzz back and forth through the cloud directing it to its destination and will have previously marked the entrance to their new home with pheromones from their Nasanov glands. Within a day the colony will have started building wax comb, using the honey reserves that they have brought with them, building down from the roof of the cavity, and as soon as comb is ready the queen will begin to lay. In another three weeks new workers will start to emerge in the new colony.

Meanwhile the old colony has been left with half of its former complement of adult bees and a number of queen cells, between ten and twenty. There will only be a small amount of unsealed brood as the queen reduced her rate of laying during the six days before the swarm left, but there will be a full complement of sealed brood and this will continue to emerge over the next 12 days and so the population of adult bees will recover to a considerable extent. From the time when the swarm left the hive, it will be another week before the virgin queens will emerge from the queen cells. During that week, the colony having hardly any brood to tend, can actively forage and, despite having a reduced population, add to the honey stores. Once the new queens emerge a number of possible scenarios will play out.

In the first of these the first virgin queen that emerges seeks out all the occupied queen cells, tears a hole in the side of the cell and stings the pupa within. The workers remove the dead queens. Should two queens emerge at the same time they will fight to the death, using their stings and attempting to maim their rival using their mandibles. If the loser is still alive the workers will complete the job by balling her, effectively suffocating her. Virgin queens, or newly mated queens, are not so easy to spot as a laying queen. Their abdomen is smaller than a laying queen, but still bigger than a worker's and often is more triangular in shape, being wider at the anterior end. They are often ignored by the workers and are less likely to be found at the centre of the brood area. Also there can be several in the hive. And being young and frisky they move quickly. But the presence of a virgin queen has the same calming effect on the colony as a laying queen and so despite there being no eggs or brood and no sign of a queen it is often possible to deduce that in all likelihood there is a virgin lurking somewhere on the grounds that the colony seems to be calm and industrious, and in particular polished cells are being prepared at the centre of the colony in apparent preparation for a newly mated queen to start laying. Much money has been wasted by beekeepers trying to introduce bought in queens into a colony which already has a virgin or newly mated queen. Usually it will not be accepted.

After a few days the queen will take a number of short orientation flights and then, when the weather is suitable, she will embark on mating flights. The mating flights require calm warm weather and usually occur in the early afternoon. Mating may be delayed until suitable conditions occur. Should she have failed to mate in 28 days after emerging, or thereabouts, she will never mate satisfactorily and will in all probability become a drone layer.

Variation in the Population of Adult Bees and Brood during Swarming

Population '000s

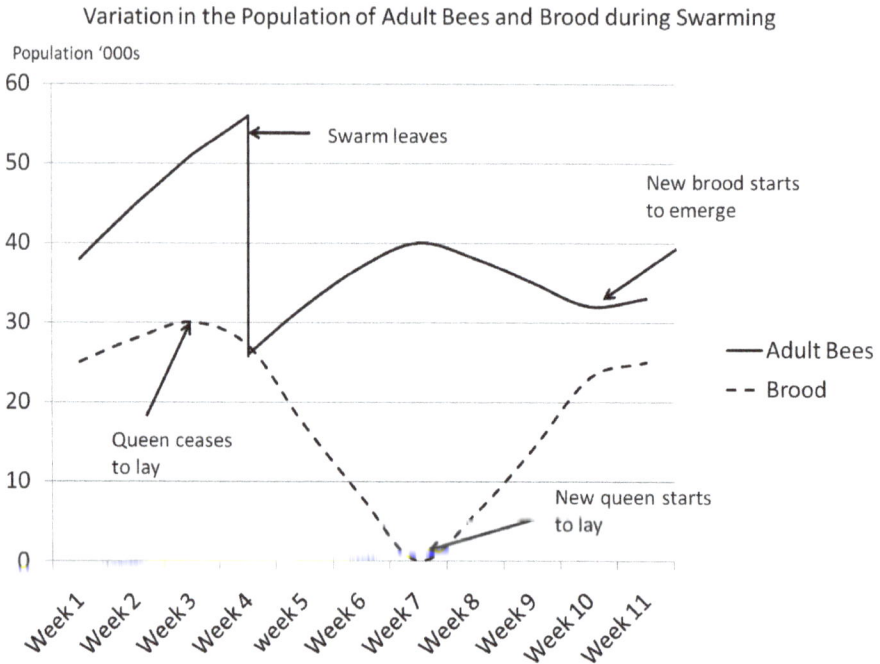

Swarm leaves

New brood starts
to emerge

Queen ceases
to lay

New queen starts
to lay

———Adult Bees

–– Brood

Week 1 Week 2 Week 3 Week 4 week 5 Week 6 Week 7 Week 8 Week 9 Week 10 Week 11

During the mating flights the queen seeks out drone congregation areas and there will mate with several drones. The mating flight may be repeated on the next day. In total she will, on average, mate with between ten and twenty drones. In the next few hours the sperm that has been deposited in her oviduct is pumped into a spermatheca, a spherical organ that can contain up to 7 million sperm. She will never mate again. The eggs that she produces over the next 3 or 4 years will be fertilized by sperm that are released from the spermatheca. A few days after mating the queen will start to lay.

Even in the most benign of conditions the new queen will not start laying until 10 days after emerging, and so there will have been a gap of 21 days during which no eggs will have been laid. Usually by the time the new queen starts laying all the brood from the old queen will have emerged. Should there have been a period of inclement weather the gap could be extended by another two weeks. It is easy to despair that a new queen will ever appear.

Afterswarms or Casts

The second scenario occurs with only the largest colonies, where afterswarms or casts are produced. Instead of killing the remaining queens the first virgin queen to emerge leads a swarm which leaves the hive three or four days after she emerges. This swarm again will take 50 – 60% of the remaining worker bees. The afterswarm, besides being smaller has a queen leading it which has not mated and therefore it will be longer before this swarm becomes established, producing its own brood. As a result, its chances of surviving the winter are generally less than for the prime swarm. Its one great advantage is that it is led by a new queen, and that new queen is the daughter of a prolific mother. Meanwhile back in the old colony the workers usually hold back the emergence of the other queens until the first afterswarm has left. When another queen emerges, she too may lead a further afterswarm, reducing the strength of the original colony even further. In the ensuing chaos some casts will have two or more virgin queens. Eventually this process comes to an end and one of the remaining queens will kill the others and establish herself as the queen of a much depleted colony. The whole process is difficult to control. The external factors that have encouraged one colony to swarm and produce casts will also cause other colonies in the apiary to act in the same way. The result can be an extremely fraught beekeeper.

For the beekeeper the whole process can be very frustrating. Strong colonies are reduced to weak ones and the promise of good honey crops come to nothing. I've visited apiaries where every other bush seemed to have swarms hanging from branches. The beekeeper can easily start running out of spare equipment. On the positive side it is an opportunity to increase the number of stocks, though the small casts will need to be established in nucleus boxes and fed generously.

Taking the Swarm

At the point when a swarm initially emerges from the hive, the swarm can be collected and placed in a nucleus or hive to form a new colony. If the swarm has attached itself to a branch at a convenient height the process can be quick and easy. It is simply necessary to hold a skep, bucket or new hive beneath the swarm, give the branch a sharp shake and the swarm will drop into the container. The top of the container can then be covered with a cloth or a lid, ensuring that the bees have ventilation. The bees can then be transferred to a permanent hive. However, it is not always so easy. There are times when the only alternative is to scoop the bees by hand into a waiting hive, or try to induce the bees in the swarm to move on to a piece of comb or a frame. This can cause a great deal of disturbance.

When the bees have swarmed on to a wall, I place a brood box containing a few frames beneath it hard up against the wall. I then set up a thin board, one edge just under the swarm and the other in the brood box, so that with a quick brushing movement, the swarm can be dislodged and will slide downwards into the brood box. The box is covered, leaving just a small entrance. These coarser methods are usually successful provided you get the queen into the box. Once she is in the box then some of the bees will start fanning at the entrance. If you look closely you can see the nasanov gland open on the dorsal side of the abdomen and the pheromones from this gland attract the remaining bees into the box.

If the swarm is high up on a post another technique is to balance a brood box with a crown board and containing a few frames on top of the post and allow the bees to move upwards on to the frames, using a little smoke if necessary. Some high tech beekeepers will turn up to collect a swarm with a gadget, like a vacuum cleaner, to suck the bees into a container.

Hiving a swarm from a skep, bucket or plastic container is usually quite easy. All that it is necessary to do is to sharply tap the container to dislodge the bees from hanging from the container sides and top, and pour them into a new hive, quickly replacing the roof. A more interesting method is to place a board sloping up to the entrance of the hive. The bees in the skep can then be poured on to the board. What you will then see is the bees, slowly to start with, moving up the board and entering the hive. In the middle of the swarm it should be possible to see the queen and so you can ensure that she too is eventually safely within the hive.

If you are wishing to hive the swarm within the apiary from which it has emerged you need to be aware that it is highly likely that the scouting for a new home was well under way, and the bees could well have been close to reaching a consensus as to where their new home should be. In which case, even though you have placed the swarm in what you consider the ideal new home, the swarming process will continue and the bees may very well abscond from their new hive and disappear off to the home that they already had in mind. This can be prevented by giving the swarm a frame of brood as bees never will abandon brood, or sealing the entrance of their new hive. Of course, if you seal the entrance there needs to be a source of ventilation, which is not a problem if you use mesh floors. Within a couple of days the swarm will have started to draw out comb and the queen will be laying. Once the swarm has made this investment it will have no desire to abscond and the entrance can be opened. Large casts, as I mentioned before, may contain more than one virgin queen and this situation can also lead to the swarm absconding. It is usually recommended that swarms are not fed straight away, but rather encouraged to use up the honey that they have brought with them in their honey stomachs from the parent hive, destroying any pathogens that they may have transferred with them. However after a couple of days, a generous feed of sugar syrup will help the new colony to become established.

There are two other situations where the bees will create queen cells.

Supersedure

When the queen is starting to fail but a swarm would be unlikely to survive then the colony will produce a new replacement queen without swarming. This process is called supersedure. There is a number of circumstances that can lead to supersedure. The colony may not be sufficiently strong to form a swarm. It may be too early or too late in the season. As this usually occurs outside the normal swarming season it is not always noticed by the beekeeper until the next season when he sees that the marked queen that he had noted in his records has been replaced by a new unmarked queen. If the colony is examined when supersedure is underway, it will be noticed that there are queen cells, but a smaller number than are normally present for swarming and these are positioned in the centre of the comb rather than at the edges. The old queen may well be still present even after the queen cell is capped, and on occasions the old and new queen will happily coexist for a few weeks, but eventually the old queen will disappear.

Emergency Queens

If a colony suddenly loses its queen it will create a small number of new queens to replace her. This process, however, relies upon the presence of eggs or newly hatched larva. The workers convert a small number of existing cells containing newly hatched larva into queen cells, refashioning their structure so that they protrude and hang down on the face of the brood comb. The larva then needs to be fed upon copious amounts of royal jelly. Experiments have shown that the later a larva is converted to being reared as a queen the less viable it will be.

Chapter 5

Pests and Diseases of the Honeybee

As with any organism the honeybee is constantly under attack from pests and diseases. These may be viruses, bacteria, fungus, mites, insects or mammals. Some of these have evolved to specifically target the honeybee. Some are extremely detrimental to the honeybees, while others are virtually harmless. At different times in history, different diseases have become the major threat. For the last twenty-five years, it is commonly accepted that the major threat is from *Varroa destructor*.

A beehive is an extremely attractive target for pathogens and pests. It is warm, protected from the elements and full of a variety of nutrition, and as a result some diseases and pests have evolved to specifically target the honeybee. As beekeepers, we may do our best to prevent our bees being infected by disease but our efforts are secondary to the natural immune systems of the honeybee. These immune systems are present at different levels. The individual honeybee is protected by antimicrobial secretions on the exterior of the exoskeleton and in the gut and should these defences be overcome, there is then the internal immune system of the honeybee. In vertebrates there are two immune mechanisms – the innate immune system and the adaptive immune system. Arthropods, including insects, only have an innate system and therefore are not able to develop immunity to new virus attacks. In many cases a viral disease can only be dealt with by introducing different genetic material into a colony by giving the colony a new queen.

But in addition, a honeybee superorganism has its own immune system, in the form of behaviour. Honeybees have behavioural patterns that ensure that the environment within the colony is kept free of pathogens. The hygienic behaviour of the honeybee results in foreign or diseased items being removed from the hive. Propolis, an antimicrobial material, is used to coat the internal surfaces of the hive and the brood cells.

Viruses

A virus is a minute entity consisting of a string of RNA within a protein shell. It has some of the properties of a living thing. In particular it can replicate itself but only within the cell of another living being, utilising the material and chemical tools within its host. In extreme cases a virus will kill the cell and may produce toxins that adversely affect the wellbeing of the entire host. A number of viruses are recognised as serious threats to the honeybee. In most cases the name gives a fairly good clue as to the effect the virus has on the bee.

Acute paralysis virus.

Chronic paralysis virus, type 1 and type 2

Deformed wing virus

Cloudy wing virus

Black queen virus

Filamentous virus

Bee virus Y

Kashmir bee virus

Sacbrood virus

In general, the viruses are not able to infect the cells of the bee without a third party vector assisting them to gain access into the tissue of the bee. The varroa and acarine mites are dangerous as they act as vectors in this way as does the nosema fungus. Early in the twentieth century honeybees in England were decimated by a disease known as the Isle of Wight disease and it was widely thought, at the time, that it was caused by acarine mite. It is now recognised that the death of the bees was due to a virus and the acarine mite acted as a vector.

Varroa Destructor

Varroa destructor is a mite. It is the size of a pinhead and can be seen easily with the naked eye. Varroa originated as a parasite of *Apis cerana*, a species of honeybee native in the Far East. Through mankind's activities it was able to transfer to *Apis mellifera* and gradually migrated west, reaching Western Europe in the 1970's and subsequently was first reported in the UK in Devon in 1992. Despite controls on the movement of bees it steadily moved north through England and by 2008 had reached northern Scotland. It was inevitable that the mite would spread, but sadly the responsibility for such a rapid movement of varroa through the country must lie with beekeepers.

It is not a successful long-term survival strategy for a parasite to kill its host. But varroa, which for millennia lived in balance with its natural host *Apis cerana*, unfortunately, when left untreated, eventually killed its Western European host, *Apis mellifera*. Beekeepers in the UK were fortunate that when their time came to deal with varroa, there were already twenty years of experience to be gleaned from beekeepers in mainland Europe. In general, in the 1990's British beekeeping continued to thrive despite the arrival of varroa, and the only ones that reached the end of their road were the beekeepers who failed to heed the advice available as to how to treat the bees for varroa. In the 1990's, treatments, based upon pyrethroids, were available. These were very successful, having an efficacy of over 99%. But during the late summer and winters of 2007 and 2008 beekeepers reported major losses. 30% to 40%, of colonies were lost throughout the UK. Though a number of factors were responsible the primary cause was judged to be varroa. For a few years prior to that time it had been recognised that the varroa mite was developing a resistance to pyrethroid and natural selection ensured that it was the pyrethroid resistant mite that was propagated. As a result, the magic bullet, the pyrethroid based treatment, no longer worked, but it needed the serious losses of 2007 to persuade beekeepers to adapt to the change of circumstances.

First, we need to understand the enemy. Varroa destructor's life cycle is entirely linked to the honeybee. The adult female varroa, during the summer, lives for two to three months during which it can have three or four breeding cycles. During this life cycle there are two stages, the phoretic stage when it is being transported on adult honeybees, and the reproductive stage inside the sealed brood cells. During the phoretic stage the mites live on the thorax, abdomen and head of the bees and feed by inserting their sharp mouthparts through the membranes joining the segments of their host's exoskeleton and ingesting the haemolymph.

During this stage they can move from bee to bee and, because of drifting and robbing, they can spread from colony to colony. During both the phoretic stage and the reproductive stage, the varroa puncture the bees' protective exoskeleton so that they can feed on the haemolymph and in this way spread viral diseases.

Timeline of varroa reproduction in drone cell

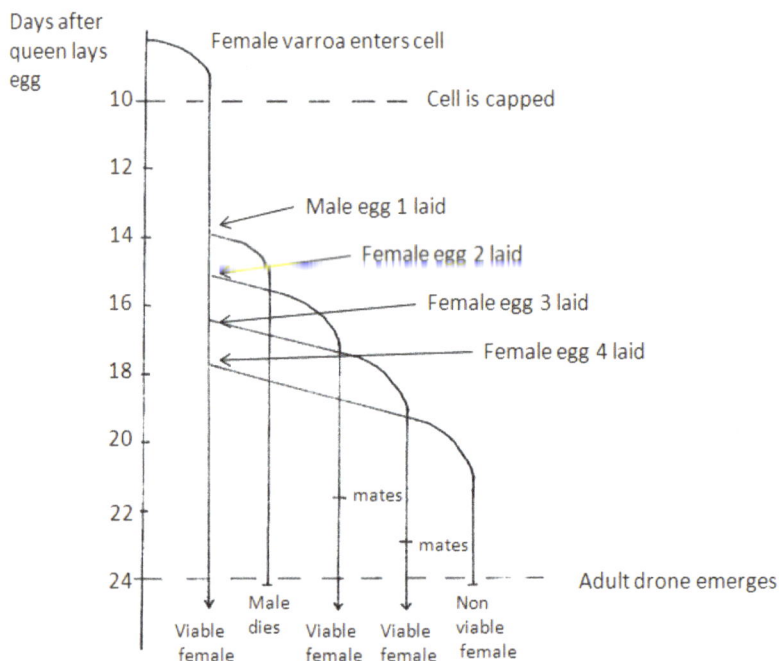

The reproductive stage starts when the female varroa moves into a brood cell shortly before it is capped. There she establishes a feeding site and 60 – 70 hours after the cell is capped she lays her first egg, followed by successive eggs at 30 hour intervals. The first egg results in a male and those that follow are all female. The development time for a female varroa nymph to reach full maturity is seven to eight days, at which point she mates with the male and emerges from the cell when the adult bee emerges. It can be seen then that the time required in the capped cell needs to be at least 11 days before the first of the next generation of mated varroa is produced. After that additional mature and fertile offspring are produced every 30 hours. We recall that workers spend about 12 days in a capped cell, while drones spend 14 days. Those extra two days that a drone pupa is within a sealed cell increases the number of viable varroa offspring that can be produced by a factor of 2. Therefore, it is not surprising that varroa favours drone brood over worker brood. In the brood cell the mites are weakening the metamorphosing bees, infecting them with viral diseases, damaging their organs and their developing wings. The male varroa and any immature females are not viable outside the brood cell and die once the cell is opened to allow the adult honeybee to emerge.

Timeline of varroa reproduction in worker cell

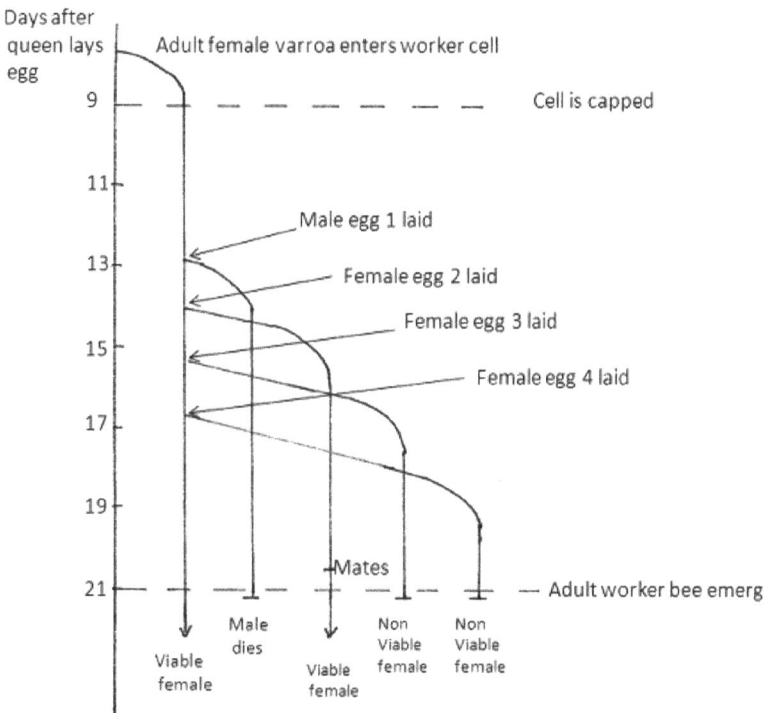

So for each cycle of reproduction the success rate is between 1 and 2 in a worker brood cell and between 2 and 3 in a drone brood cell. Bearing in mind that there can be at least eight cycles of reproduction during the summer and taking a conservative average of 2 for the reproductive rate for each cycle, it can be seen that the population will increase by 256 times (2^8) during the course of a summer. This is an approximation for the purposes of illustration and some sources suggest that the potential growth of the varroa population can exceed this. But even taking this conservative annual rate of reproduction it can be seen that a treatment which has a 99% efficacy applied once a year is not actually sufficient to prevent a long-term growth in the population of varroa in a hive.

It can be seen that varroa is a formidable pest. It weakens both the adult and larva honeybee and spreads viral diseases. The original host species *Apis cerana* had a shorter brood period and so the reproductive success rate of the varroa was less and *Apis cerana* does appear to have evolved a grooming behaviour which dislodges the mite from the bodies of the bees during the phoretic stage. *Apis mellifera* does not have these advantages. If a single adult female varroa is introduced into a colony of *Apis mellifera* then, without intervention, in two to three years that colony will have become so infested that it will collapse. And we can be reasonably confident that every colony in the UK is infested to some degree. I don't want to be too negative. Beekeepers have now learnt how to live with varroa and during the period since varroa has become widespread throughout the UK, the number of beekeepers and the number of hives that they are maintaining has increased significantly. The real toll has been in numbers of feral colonies in our towns and countryside. They are not receiving treatment and accepted wisdom is that most of them will have become infected and eventually succumb.

Integrated Pest Management

When varroa destructor first arrived, colonies were treated once a year in the early autumn with plastic strips impregnated with pyrethroids. Initially the treatment was very effective and would kill over 99% of all mites. It was realised at an early stage that the mites would inevitably develop immunity to the active pyrethroid ingredient and by 2004 – 2005 resistant mites were being detected, apparently spontaneously arising in many parts of the country. There is no other single treatment available that is as effective as the pyrethroid strips and so beekeepers have been forced to adopt the more sophisticated strategy of integrated pest management (IPM).

IPM is a concept taken from established practice in agriculture and horticulture. Within IPM there is the assumption that it is not possible to totally eradicate a given pest. The aim of the strategy is to maintain the population of the pest below a level where it adversely affects the crop that is being protected, in our case the honeybee. An IPM strategy generally will use a combination of weapons at different times of the year to combat the pest, selected from the available armoury. So what weapons are available to protect the honeybee from varroa?

Chemical Varroacides

Where an authorised, proprietary medicine is used it is essential that the instructions are read and understood. Besides the method of application, note needs to be taken of recommended storage, use by dates, protective clothing required, and methods of disposal etc. Do not be misled into believing that natural substances are necessarily safe. Indeed, some of the natural acids can be extremely hazardous, and it is only sensible that beekeepers give their own wellbeing an even higher priority than that of the bees. Under no circumstances should beekeepers be tempted to use agrochemicals which were developed to control pests on other animals. They can be dangerous and the beekeeper would be liable to legal action if residues of unauthorised substances are found in their honey.

Beekeepers are required by the Veterinary Medicines Regulations 2013 to maintain a register of the chemicals used in each hive, including the name and batch numbers of products used, the date of administration and the quantity administered. Further details can be found on BeeBase.

Pyrethroids

Pyrethroids are sold under the trade names Apistan and Bayvarol. The product consists of plastic strips impregnated with pyrethroid. The strips are suspended between the brood frames and left in place for four weeks and then removed. When varroa destructor first arrived in the UK the treatment was very effective and would kill 99% of all mites, and this was sufficiently effective to allow the colony to survive for another year without further treatment. However now that the pyrethroid resistant mite is widespread across the UK it is no longer a sensible option to use.

At the same time, it has been discovered that residues of pyrethroids are retained in wax, to the extent that most purchased foundation is contaminated to some extent. There is some evidence that pyrethroids are responsible for reducing the fertility of drones, and it is speculated that this has been a contributory cause of the poor results of queen mating in recent years.

Thymol

Thymol has a long history of use against varroa. If used in accordance with instructions accompanying the products there appear to be no adverse side effects on the bees, adult or brood, and, besides, thymol occurs naturally in honey in small quantities. However, the efficacy of thymol based treatment is, at best, about 90%, and it needs to be applied when ambient air temperatures are above 15°C.

There are a number of approved products, registered as medications for honeybees, where the principle active ingredient is thymol. It is possible to use thymol crystals directly but any beekeeper adopting this form of treatment needs to be aware that thymol in its concentrated form is a hazardous substance.

Apiguard is a gel containing thymol, which is designed to allow the thymol to vaporise at a slow rate over a period of about a fortnight. Beekeepers can buy the product in sealed trays that represent a single dose. A full treatment requires two doses. The four week treatment period is required because of the life cycle of the varroa mite and thymol is only effective on varroa during its phoretic stage. For the twelve to fourteen day period that the mite is within a sealed cell it is protected from thymol treatment. Apiguard can be applied in the spring or the late summer, before the daytime temperatures regularly drop below 15°C, which normally happens towards the end of September. So, it can be deduced that the Apiguard treatment needs to be commenced before the start of the final week of August. Also, it needs to be born in mind that it is not really possible to successfully do autumn feeding while still treating with Apiguard, and the feeding needs to be completed by the end of September. Even though the active ingredient can be found naturally in honey the treatment should not be used when there is honey on the hive that is intended for harvest and sale.

Varroa mite
(Courtesy the food and
environment research agency (FERA),
Crown copyright)

Sealed drone brood being tested for varroa
– positive result
(Courtesy the food and environment research
agency (FERA), Crown copyright)

Colony being treated with a tray of Apiguard. Note eke to make the space to allow the bees to access tray.

The trays are opened by tearing away the top foil and then the tray is placed directly on to the brood frames. There is no need to touch the gel, but gloves should be worn. An eke (an eke is described in detail in the next chapter) should be placed about the tray to create the space for the bees to access the paste. Alternatively crown boards can be modified by attaching battens to the edges of one side to create the space. The thymol vapour is heavier than air and so if a mesh floor is being used the sampling board must be inserted and sealed in place with duct tape. The warmth and activity of the bees causes the thymol in the gel to vaporise and this affects the cell membranes of the mites, killing them. After two weeks a second tray should be put into place. Any remaining gel in the first tray can be scraped on to the top of the brood frames. After another two weeks the second tray can be removed.

The manufacturers claim that Apiguard is 93% effective, but the efficacy will depend upon the ambient temperature. At the present time there are no indications that varroa is forming any resistance to treatment by thymol. One of the disadvantages of Apiguard is the space that needs to be formed above the tray, which the bees of a strong colony will fill with wild comb. At the end of the treatment this wild comb must be removed, impossible to do without upsetting the bees and causing collateral damage.

ApiLifeVar is another product whose main active ingredient is thymol, though it also contains eucalyptus oil, menthol and camphor. These ingredients are absorbed into an inert medium forming wafer like bars. Each application consists of a bar being broken in four and the quarters being distributed across the top of the brood frames. The total treatment consists of two to four applications at intervals of 7 – 10 days. Its advantage over Apiguard is that there is no requirement to create a space for a tray. The manufacturers claim an efficacy of about 90%. As the treatment time is shorter than Apiguard, the treatment can be started rather later, maybe during the first week of September.

Using a single annual treatment of Apiguard or ApiLifeVar is not sufficient to control varroa, but both are important tools in an IPM strategy.

Oxalic Acid

Oxalic acid is cheap and effective. It is can be bought in the form of crystals which is not an approved medication or in a diluted form, two forms of which are now approved, Oxybee and Api-bioxal. Oxalic acid in its concentrated form is a hazardous substance, so beekeepers are recommended to buy it in diluted form, normally between 3% and 6%, usually combined with sugar solution. It appears to be harmless to adult honeybees but can have an adverse effect on the brood of the honeybee. There is an ongoing debate about the sub-lethal effects of oxalic acid and other treatments. Also, the treatment is only effective against varroa in their phoretic stage. Therefore, oxalic acid is only applied when a colony is brood less. As a result, oxalic acid is applied either during the middle of the winter when there is no brood, when a swarm is newly collected or as a part of the shook swarm procedure. A dose of the solution is applied by trickling 5ml from a syringe or a dispenser between each pair of combs and on to the cluster of bees.

Oxalic acid can also be used in crystal form and vaporised within the brood box. Equipment, powered by a 12 volt battery, is available for this. A warming iron with a spoon shaped end, on which is placed the oxalic acid crystals, is inserted through the entrance of the hive. The crystals are heated so that they sublimate into a vapour within the hive. The hive entrance needs to be sealed but the procedure takes just a few minutes for each hive. This can be very effective but the oxalic acid fumes are potentially very dangerous to the beekeeper and so this procedure should only be used if the beekeeper is equipped with the appropriate protective clothing and masks. Oxalic acid treatment, when used as described above, can have an efficacy of 90%, and once again it is an important weapon to be used as part of IPM.

Formic acid

Formic acid is a very effective treatment, killing all forms of mites, acarine as well as varroa. In contrast to other treatments it will penetrate sealed cells and so a single treatment is sufficient. Formic acid is potentially a very hazardous substance and safety and administration instructions must be followed carefully when using it. Its efficacy depends on temperature and it is best used in late spring, or in early autumn after the honey supers are removed. As formic acid is naturally found in honey, it does not contaminate the honey. MAQS is an approved treatment where formic acid is the active substance. The treatment consists of pads impregnated with formic acid which are laid on top of the brood frames, minimising risk and inconvenience. The pads are just 6mm thick and so fit in the bee space between boxes, eliminating the need for ekes. Also, the recommended treatment time is much shorter than most of the thymol based treatments. As with all things that sound too good to be true, there is a big down side. The fumes from the MAQS pads are extremely pungent and do appear to cause considerable agitation to the colony and there is some question as to whether the treatment is suitable for a weak colony. Formic acid may corrode iron parts in the hive, such as mesh floors, castellated spacers and even the nails.

Other Essential Acids

There are other substances that have been used at one time or another to treat varroa, such as lactic acid and various essential oils. In the UK these are rarely used and there are no commercially available products to apply these substances. They are not authorised medicines and there are no generally recognised dosages or methods of applications.

Biotechnical Methods

There are number of strategies that can be used that reduce the varroa population which do not involve the use of chemicals.

Drone Brood Removal

Because the time for which the larva is within a sealed cell is longer for drones than for workers, varroa have evolved to prefer to breed in drone cells. This preference can be exploited to control the varroa population, by selectively removing drone brood. If a super frame with foundation is inserted into a brood box the bees will draw out drone brood comb beneath the bottom bar. Alternatively, some beekeepers insert frames containing drone foundation. When the drone comb is sealed, which would be between two and three weeks after the frame is inserted, the drone comb can be cut away and destroyed, and the process repeated. Any varroa mites that had chosen to breed in the comb are then destroyed. The procedure must be repeated throughout the swarming season. This strategy certainly controls the varroa population but at a cost to the colony. And it requires discipline on the part of the beekeeper because if the beekeeper fails to remove the drone comb within 24 days after the eggs were laid the drones will emerge and with them will emerge a new generation of varroa, exacerbating the situation. Another downside is that the honeybees in the apiary and the surrounding area are being deprived of drones. Drones have a vital role to play in the life cycle of the honeybee – they mate with the virgin queens. Reducing the drone population reduces the chance of the queens mating successfully, and indeed a major problem recently has been the failure of virgin queens to mate sufficiently well so that they are able to continue to lay for three or even four years.

Mesh Floors

Fitting mesh floors, without the sampling tray, allows mites that fall from the backs of bees to fall through the mesh on to the ground, from where they are unable to return to their hosts. When the sampling tray is put in place it is possible to monitor the number of live mites that fall through the mesh. The numbers of mites that perish in this way is significant but nowhere near enough for this procedure to control varroa if used alone. However, it can be used as a part of an IPM strategy.

Icing Sugar

By sprinkling icing sugar on the combs, the bees are forced into a grooming behaviour that dislodges the mites and at the same time the powdery substance on the bodies of the honeybees gives a poorer foothold for the mites. The result is that a percentage of the phoretic mites will drop to the floor of the hive, and if a mesh floor is fitted, then on to the ground. The easiest and quickest method to apply icing sugar is to place mesh over the brood frames and sprinkle 100g of the icing sugar on to it. Then, using a bee brush, the icing sugar should be brushed through the mesh, distributing it on to the top of the brood frames. The efficacy is low but as the sugar is harmless and the treatment is quick and inexpensive, this procedure can be carried out with each inspection during the summer months. Again, the procedure will not control varroa when used alone, but is another weapon in the armoury and can be used as part of IPM.

Bee Gym

Recently a gadget has come on the market, called a bee gym, which is designed to encourage bees to groom themselves, and in so doing dislodge the mites on their bodies. If the bee gym is placed on the mesh of an open mesh floor the dislodged mites will fall to the ground.

Shook Swarm Technique

The shook swarm is a procedure that is used to renew brood comb when a colony is well established in spring. As a consequence of the procedure all brood is destroyed, and so all varroa that is within the sealed brood cells is also killed, and the short period after the procedure is completed, when there is no brood, is an ideal time to treat with oxalic acid. The shook swarm procedure is described in detail in the section on comb renewal in chapter 12 – Keeping Healthy Bees. I would suggest that the shook swarm procedure is one of the most effective treatments for varroa.

Integrated Treatment for Varroa

During the peak of the summer the varroa population can be quite large, without significantly affecting the well being of the colony. For example, a varroa population of 5000 mites would only be directly affecting one in ten of the workers. But as we go into August and then September the population of bees decreases while the population of varroa, slowly and inexorably increases so, within a matter of weeks, it is possible to reach a situation where one in three (or worse) of the workers are directly affected. A colony is not able to tolerate this level of infestation and would collapse. The collapse of a strong colony from varroa can happen remarkably quickly.

One of the most critical times of the year for the honeybee colony is the late summer. If serious problems occur at that time there is little chance for the colony to recover, even with intervention from the beekeeper. During August, September and October the colony is producing the winter bees, the bees that will form the winter cluster and form the work force when the colony starts to increase in the following spring. Whereas the summer bees may only live five or six weeks as adults, the winter bees are required to live five or six months. These bees need to be healthy and well nourished.

As has been said, varroa are not simply dangerous because of the physical damage that they do to both the adult bees and the brood, but also because of their role in spreading viral disease, in particular acute bee paralysis virus and deformed wing virus. This needs to be borne in mind when deciding upon your IPM strategy to control varroa. At the same time, treatments must avoid contaminating honey that will be extracted for human consumption.

The IPM strategy that I now use, and this is after experimenting with alternatives, is

a)   Use mesh floors all the year round

b)   Apply icing sugar to the brood frames as part of my regular inspections during the swarming season

c)   Treat with Apiguard or MAQS in the middle of August through to the middle of September

d)   Treat with oxalic acid by the trickle method in December

By applying this strategy, I believe I am keeping the varroa mite at low levels, so that their adverse effect on the well being of the colony is minimal. This is not a definitive solution and I know that other beekeepers are keeping their bees successfully with different strategies to control varroa. The important point to remember is that a single procedure or treatment is unlikely to be sufficient.

Monitoring the Varroa Population

It is not sensible to wait to see whether or not your colony collapses before judging on the efficacy of your strategy. A better policy is to periodically monitor the level of varroa mite infestation. There are two principle ways to do this.

The first method is by monitoring the natural mortality of varroa. This can be done by inserting a sampling tray beneath a mesh floor for a fixed period, say two or three days, and counting the number of varroa mites that fall through the mesh floor on to the tray. At the end of the period the number of mites on the tray should be counted and the count converted into 'daily mite drop'. Where there is other debris and it is difficult to pick out the mites, one solution is to put the debris into a jar of methylated spirits and the mites will become more evident as they float. The natural mite drop in a colony is related to the size of the varroa population. Research suggests that the colony may collapse before the end of the year if

a)   In spring the daily mite drop exceeds 0.5
b)   In May the daily mite drop exceeds 5
c)   In June the daily mite drop exceeds 5
d)   In July the daily mite drop exceeds 6
e)   In August the daily mite drop exceeds 7
f)   In September the daily mite drop exceeds 7

The second method is by uncapping drone brood. As has been explained before, varroa mites favour drone brood to breed in. An area of drone brood that has been sealed for a week should be selected and then using an uncapping fork, the prongs should be inserted, parallel to the surface of the comb, under the domed cappings. The pupae can then be lifted out and examined. The varroa, if they are present, can be clearly seen, dark dots on the creamy white flesh of the pupae. About a hundred pupae should be examined and the proportion that have been infested with varroa estimated. If more than five percent of the drone brood cells have been infested then the colony is at risk.

In the past the effect of an epidemic affecting honeybees has decreased with time, either as the honeybee adapts its physiology or behaviour to successfully combat it, or as the disease or pest evolves to become less virulent and less deadly. At the present we are still waiting to see how the situation will resolve itself. A number of research projects are underway to selectively breed bees that have grooming or hygienic behaviour patterns which target the varroa mite. It could be that natural selection in the feral population will result in a strain of bees that can coexist with varroa. It has always been expected that feral colonies will eventually succumb to varroa, but occasionally one hears reports of feral colonies that appear to be defying this general prognosis. An alternative view is that a longer-term solution will arise when the varroa mite evolves to become less virulent. A parasite that kills its host is not a viable entity. It would be a logical outcome of natural selection that, where the more deadly strains of varroa will become genetic cul de sacs, strains of varroa that can live in balance with the European honeybee, as they do with the eastern honeybee, will survive. The life cycle of varroa is much shorter than that of the honeybee so it is more likely that varroa will adapt than the honeybee. Time will tell. In the immediate future we must take the necessary steps to enable our bees to survive the presence of the varroa mite as it is now.

Brood Diseases

This next section looks at other diseases of the brood. Two of these are known as foul brood. They tend to be linked together but are in reality very different. Their names are European foul brood (EFB) and American foul brood (AFB). Both are subject to statutory controls in the UK, the Bee Diseases and Pests Control (England) Order 2006 No 342, with separate orders in Wales, Scotland and Northern Ireland. As a result, beekeepers are legally required to inform the National Bee Unit should they suspect that their bees are infected with either of these conditions, and FERA is empowered to take any measures necessary to control the disease. If EFB or AFB is suspected the beekeeper should impose a voluntary standstill on movement in and out of the infected apiary. The entrance of the hive should be reduced but not closed. The apiary will be inspected by bee inspectors as soon as possible and if the disease is confirmed a statutory standstill notice will be issued prohibiting movement of all honeybees in the apiary. The bee inspectors will take the appropriate measures to eradicate the disease. This will include inspecting all colonies in other apiaries in the immediate area.

European Foul Brood

Despite the name this disease is found throughout the world, wherever there are honeybees. It is a disease of the larvae of the honeybee. The pathogen is a bacterium called *Melissococcus plutonius* and infects the mid-gut of the larvae. It is thought that the disease is endemic in the UK and outbreaks of EFB tend to occur in spring and other times when colonies are under stress, for instance during periods of poor weather. But outbreaks of the disease are relatively rare. In 2009, in Yorkshire, the bee inspectors inspected over 3324 colonies and came across 52 cases of EFB. The overall incidence of the disease is lower than these figures indicate as the bee inspectors will be called in to inspect suspect colonies. The incidences are at historically low levels, probably due to the vigilance of the bee inspectors, better methods of detection and the ever-improving education provided by the beekeeping associations. The bacteria multiply in the gut and compete with larva for the food, so that larvae that succumb to the disease are effectively starving to death. Larvae that are adequately fed may occasionally survive and it is possible for a strong colony to naturally recover from a mild outbreak, but the spores of the bacterium will remain in the brood comb and this will lead to the risk of further outbreaks in the future. The bacteria are in the faeces voided by an infected larva, and then are spread by the workers as they clean the brood cells. The disease is very contagious and is readily spread throughout the colony, from colony to colony and apiary to apiary. Serious outbreaks will result in the demise of the colony.

When checking for foul brood, alarm bells should ring if the appearance of uncapped larvae diverges from the normal appearance of healthy brood, the pearly white colour, clearly segmented and with the larva curled in the base of the cell in a 'C' shape. The infection causes the larva to become distorted and lie in an unnatural attitude. The gut of the larva can become visible through the translucent body wall. And then when a larva dies it collapses as though it has melted. Some larvae die after the cell has been sealed, and this results in sunken, perforated caps. The worker honeybees remove the dead and dying larvae and as a result the brood comb takes on a patchy, pepperpot appearance. Though the pepperpot appearance is an important symptom, it is not definitive.

Brood infected with EFB
(Courtesy the food and environment research agency (FERA), Crown copyright)

Brood infected with EFB. Note distorted position of some of the larvae
(Courtesy the food and environment research agency (FERA), Crown copyright)

Ultimately the definitive diagnosis will be made by the bee inspector. There are now lateral flow devices that can give an immediate confirmation. There are a number of actions that a bee inspector may take. If the infection is severe and there is little chance of the colony recovering then the colony will be destroyed. The dead bees, brood frames and comb are burnt in a pit and buried and the hive boxes sterilised with a blow torch. For less severe outbreaks the colony will be put on to fresh comb using the shook swarm procedure, or treated with antibiotics. Shook swarm is now the preferred treatment. The comb removed during the shook swarm procedure is also burnt. The standstill order will remain in place for several months during which the colonies will be re-examined by the bee inspectors to ensure there is no re-infection.

Beekeepers must take responsibility for preventing the spread of this very infectious disease, as it is beekeepers that have been shown to be the main vectors for the spread of the disease. When you bring bees or equipment into your apiary you may well be also bringing in disease.

Second hand equipment should be sterilised and secondhand frames and comb destroyed. Swarms and new colonies need to be put into a quarantine apiary until they are well established and you have carried out an inspection of the brood.

American Foul Brood

Again, the name is unhelpful, as American Foul Brood is not confined or ever was confined to America. The pathogen is a bacterium called *Paenibacillus larvae*. Once a colony is infected it will inevitably die. The bacteria produces spores that are extremely resistant to heat and chemical treatment and can survive for up to 40 years. Fortunately, the eradication procedures put in place by the NBU limit the number of outbreaks. In 2009 in Yorkshire there were 17 cases and that was a record high for recent times.

If the larva consumes food contaminated by the spores of *Paenibacillus larvae*, the spores germinate in the gut and move into the tissues of the larva where they multiply at an enormous rate, consuming the tissue. The larva normally dies after the cell is capped, unlike EFB where the larva normally dies before the cell is sealed. The remains of the pupa are highly infective. The cell caps become sunken, perforated and damp looking and the pupal remains are brown with a slimy consistency. A definitive diagnostic method is to insert a matchstick into a suspected infected cell and withdraw it. If the remains can be drawn out in a brown, mucus-like thread up to 25mm long, this is a reliable confirmation of AFB. This is known as the 'ropiness' test. To start with the workers will endeavour to clear infected cells, and this results in a pepperpot appearance of the brood. The bee inspectors have lateral flow devices to confirm the diagnosis. Unfortunately, the workers' attempts to clear the infected cells results in the disease spreading further and the colony soon collapses. Dead pupa left in the cells dry to form a dark scale on the lower side of the cell, and these are difficult to remove. The proboscis of the dead pupae sometimes remains intact and can be seen protruding from the scale.

Brood infected with AFB. Note slimy appearance of the wax cappings
(Courtesy the food and environment research agency (FERA), Crown copyright)

Carrying out the ropiness test to test for AFB
(Courtesy the food and environment research agency (FERA), Crown copyright)

If a beekeeper suspects that a colony has AFB, all movement in and out of the apiary must cease and the NBU should be informed. Once an AFB infection is confirmed a statutory standstill notice will be served and the colony destroyed. The bees are killed and the frames and comb burnt in a pit and the ashes buried. The remaining woodwork of the floor, brood and super boxes can be sterilised using the flame of a powerful blowtorch. I think I would be inclined to burn the boxes as well.

Though the incidence of AFB is now rare its effect can be devastating on the beekeepers concerned, losing not just their bees but parts of their equipment as well. Some beekeepers take out insurance, either individually or through their beekeeping association, with 'Bee Diseases Insurance Ltd' which will compensate against losses brought about by foul brood diseases. BDI also promotes and sponsors research into foul brood diseases.

Nosema

Nosema is a single celled microsporidian, a fungal type organism, which multiplies in the ventriculus, the large intestine of the honeybee. This affects the ability of the honeybee to digest pollen and sugars, and so weakens the bee, reducing its ability to produce brood food and wax and shortens its life. The cumulative effect of thousands of weakened bees is to slow down the rate at which the colony can build up.

There are two types of nosema, *Nosema apis* and *Nosema cerana*. *Nosema apis* has long been known as a disease of *Apis mellifera*, but in recent years *Nosema cerana* has crossed species from *Apis cerana*, the eastern honeybees, to *Apis mellifera*. Both species of nosema can be detected by a microscopic examination of the contents of the bees gut, but it is not easy to absolutely differentiate between the two without an analysis of the genetic makeup. *Nosema cerana* is now widespread throughout Europe and there are fears that its effect on honeybees is more serious than *Nosema apis*. There are no longer any approved medicinal treatments for Nosema.

The spores are in the faeces voided by the bees and through the faeces the disease is spread. However, it will be remembered that honeybees normally void their rectum outside the hive. In the winter the honeybees allow waste material to accumulate in the rectum which expands within the abdomen. During the course of the winter, on sunny days, bees leave the colony for short cleansing flights to void this faecal matter. In these circumstances the nosema is not readily spread. The problem occurs when there are long periods of cold or inclement weather and the bees are unable to leave the hive. Another risk situation is when the bees have dysentery. This can be caused by the winter stores having excess water, and wintering on heather honey is thought, on occasions, to be a cause. The dysentery results in the bees voiding within the hive, leaving brown stains on the frames and at the hive entrance.

If nosema is diagnosed during the spring, a renewal of combs as part of the spring inspection, using Bailey frame change procedures, reduces the reservoirs of nosema spores in the colony and the natural growth of the colony and the behaviour patterns of the bees during the spring and summer months will allow the colony to recover.

Acarine

Acarine is disease caused by a mite, *Acarapis woodii*. At one time it was thought to be the causative agent of the Isle of Wight disease which killed the majority of honeybees in the UK during the first part of the twentieth century. The mite infects the trachea of the honeybee, using the tracheal sacs as sites for reproduction and by inserting their mouthparts through the cuticle within the trachea, feeding off the haemolymph of the host bee. The mature female mites, having mated, leave the trachea via the spiracles and attach themselves to the hairs of the thorax and when the opportunity arises transfer themselves to young bees that are less than nine days old and move into their trachea where 5-7 eggs are laid. The young mites are mature 14 days after the egg was laid. There are few serious symptoms directly resulting from an acarine infection, though acarine must inevitably reduce the vitality of the individual bees and therefore of the colony as a whole. But more significantly, like varroa, the mites may act as vectors for the spread of viral disease, in particular the chronic bee paralysis virus.

Acarine can be diagnosed by dissection of the honeybee, removing the head and the thoracic collar and examining the thoracic trachea with a low power dissecting microscope or even a strong magnifying glass. The trachea of an infected bee will be discoloured with brown areas, lacking the creamy white appearance of a healthy bee.

There are no registered treatments for acarine, but the chemical treatments that are used for varroa mite are believed to also have an effect on the acarine mite. Probably because of the widespread use of the treatments to control the varroa mites, acarine is not a serious problem at the present time.

Wax Moth

There are two species of wax moth that are of concern to the beekeeper – the lesser wax moth (*Achroia grisella*) and the greater wax moth (*Galleria mellonella*). In both cases the moths breed within the honeybee colony, the larvae feeding on beeswax, pollen, honey and hive debris. Strong colonies give the wax moth short shrift, killing and driving away invading moths and ejecting the wax moth eggs and larvae. On the other hand, weak colonies can be quickly overwhelmed, with the larvae burrowing through the comb, exposing the honeybee pupae and spoiling the honey and eventually totally devouring the wax comb. Nevertheless, if a holistic view is taken, the wax moths have a useful part to play in the health and well being of honeybees. By invading weak diseased colonies and then clearing out the wax combs they destroy what otherwise would be reservoirs of pathogens and restore the cavities to a condition where they can be reused by swarms in subsequent seasons without fear of re-infection.

The lesser wax moth is widespread throughout the UK, but the greater wax moth was, until recently, rare in the north of England, though not unknown. It is feared that, in the future, global warming will extend the range of the greater wax moth further north. It is the greater wax moth that causes the most damage, not just to the comb but also to the woodwork of the brood and super boxes and frames. The larvae burrow out indentations in the wood before forming their cocoons, and in extreme cases this can compromise the integrity of the hive and frames.

The best method of deterring wax moths is to keep strong colonies. The best encouragements to wax moth is having failing colonies and leaving used brood comb in an empty hive during the summer months.

The wax moths seem to like dark corners for laying eggs and forming their cocoons and so the use of mesh floors is a discouragement though not an absolute deterrent. To keep stored comb free of wax moth, at one time we were advised to use paradichlorobenzene crystals placed on sheets of newspaper in the stacks of brood boxes and supers containing wax combs. This is no longer approved as PDB is carcinogenic and contaminates the wax. There is a biological agent available to control the wax moth larvae. But in my experience it is possible to control wax moth in stored comb, by only storing wax combs that have contained honey, keeping them in a cool place and stacking the boxes so that they can be sealed. Old brood comb should be rendered down at the first opportunity.

Chalkbrood

Chalkbrood is a common but not particularly serious disease. It is caused by a fungus, *Ascosphaera apis* which invades the body of a developing larva and eventually kills them. The worker bees remove the cappings to expose the bodies of the larva which appear as white, hard mummies. The mummies shrink and are easily removed from the cell and can be found on the hive floor or on the ground close to the entrance. It is a disease that is associated with damp weather in the spring and as the summer approaches the incidence of the disease will reduce. Chalkbrood is less likely to occur where the colony is maintaining the full brood nest temperature of 34°C to 35°C, and so is more common at the extremities of the brood nest or in small colonies that are struggling to maintain the brood nest temperature. If a colony is particularly prone to the disease it is sometimes recommended that the colony is requeened. It is a disease that comes and goes without having any major effect on the wellbeing of the colony.

Other Pests

Mice can be a problem to honeybees during the winter. To mice the hive can appear to be a warm, dry refuge in which to spend the winter, with a good supply of food available. Once the bees are in a cluster there are no guard bees to prevent the mice gaining entry. Though it is possible, for a while, for a nest of mice to coexist with the cluster, the mice damage sections of the comb, damage the frames and even the brood box. On some occasions the colony will be affected to such an extent that it will fail to survive the winter. To overcome the risk from mice, the entrance needs to be reduced so that it is too small for mice to enter. There are two approaches to this. Mouse guards can be made or purchased. These consist of strips of galvanised metal with a mesh of 9mm holes punched into them. At the beginning of the winter these strips can be tacked in front of the entrance. The 9mm holes are just too small to allow a mouse's head to squeeze through. But now beekeepers are using mesh floors it is possible to use a permanently reduced entrance slot, 100mm by 8mm. Again, the 8mm width is too narrow for the mice to squeeze through.

Woodpeckers

Green woodpeckers can also be a real nuisance during the winter. Many apiaries are never bothered, but once a population of woodpeckers learns that there is a significant source of insect food within hives, which to them are no more than stumpy trees, then they will return again and again. Not only do they predate upon the bees, risking the long-term survival of the colony, but they damage the brood boxes, sometimes beyond repair. The best method of protection is by wrapping the hives in chicken netting, inserting battens between the netting and the hive body to prevent the woodpeckers being able to peck the wood of the hive. Care needs to be taken not to obstruct the hive entrance. If there is a history of woodpecker damage in the area then this precaution needs to be taken as a matter of course as part of winter preparations. There is no sense in waiting to see what happens before acting. There are alternative methods which beekeepers use to deter woodpeckers, such as pinning or hanging old cd's from the hive. This may act as a deterrent in the first place but once the woodpeckers have discovered the goodies inside a hive they are as good as useless.

Wasps

Wasps are in need of much improved public relations. Nobody has a good word to say for them. In actual fact, during the spring and the first part of the summer they are a useful part of a balanced ecology. They prey on aphids and by so doing they are a real boon to the gardener. The aphids are taken back to the wasp's nest and are used to feed the larvae, and as a by-product a sweet solution is produced that the worker wasps feed upon. The problem arises in August when the colony has produced the new queens and drones and old queen ceases to lay. The workers then have no source of sugar from the brood food and start to search the surrounding area for alternative sources of sugar. The honeybee colony then becomes an obvious and inviting target.

A strong colony of bees in a well maintained hive is under no real threat. The guard bees are able to defend the colony, especially if the beekeeper helps by reducing the entrance. Weak colonies, on the other hand, can collapse under the sustained attack of wasps. Once a wasp manages to breach the defences at the entrance to the colony, it is able to wander about within the hive, apparently unchallenged.

The beekeeper can take a number of steps to pre-empt and reduce the threat. It is important to maintain high apiary hygiene standards, making sure that there is no old comb lying about on the grass or spills of sugar solution, nothing to attract the stray wasp. Direct action can be taken by setting up wasp traps close to but not in the apiary. In their simplest form they consist of a one pound jar, half filled with flavoured, sweetened water, and a 6mm hole drilled in the lid. The wasps squeeze through the hole into the jar but are unable to get out. After a few days you have a jar full of dying wasps. Wasp nests can sometimes be located by following the wasp workers and then the nest can be killed using proprietary powders. The wasp threat varies from year to year. Some researchers have suggested that wasp populations wax and wane on a seven year cycle.

Asian Hornet

It is feared that the Asian Hornet, *Vespa Velutina*, could be the next major threat to honey bees in the UK. The Asian Hornet was first observed in south-west France in 2004 and in the following decade it extended its range to cover most of France. The first sighting of an Asian hornet in the UK was in 2016 in Gloucestershire, and there have been several others in subsequent years. In each case, so far, the NBU believe they have eliminated the incursion.

The Asian Hornet worker is about 25mm in length, somewhat smaller than the native hornet. The queen measures about 30mm. The abdomen is mostly black, except for the fourth segment which is yellow. It has yellow legs and an orange face. The life cycle is similar to a wasp. In the spring the queen emerges from hibernation and builds an embryonic nest. Later the colony usually relocates and develops a new nest, often high in a tree, hidden by foliage from viewers at ground level.

During the summer the population in a nest may reach 6000 individuals. The Asian hornets predate on insects and in particular on honey bee colonies. They take part in hawking behaviour, hovering outside hives waiting for bees to return. The bees are caught and the protein rich thorax is taken back to their nest and fed to their larva. This behaviour can cause significant damage to honeybee colonies.

The priority at the present is to monitor incursions of the Asian hornet from continental Europe and destroy any colonies that are found. Whether this policy will be sustainable in the long term remains to be seen.

The Asian hornet has had a devastating effect on beekeeping in France. However, we will be able to benefit from their experience and there do seem to be methods of managing our bees that should mitigate the possible damage. Wire mesh frames can be attached in front of the hive entrances and French beekeepers have learnt that it is necessary to feed both sugar and pollen substitute during August and September when the hornets' predation is at its worst.

Exotic Pests

The adjective 'exotic' in this context hasn't the same connotations as when qualifying the noun 'dancing'. Except, that is, that I have no personal knowledge of either. Exotic pests of honeybees are pests of the honeybee in other countries, but not so far, in the UK. The greatest threat and one which we are very likely going to need to confront in the future is the small hive beetle. This is a native of Africa but is now endemic in America where it is the cause of large losses of honeybees. The beetles lay eggs in the hive and the larvae feed on the wax, pollen and honey causing great damage. The larvae, when they are fully grown, leave the hive and pupate in the ground close to the hive. The adults can fly and so can quickly infect neighbouring colonies, and it is believed that they are attracted to hives by the smell of the pheromones produced by the bees. As the beetle is a native of Africa it is not clear whether the small hive beetle will survive the British winter. The informed suspicion is that it will be able to and so determined steps are being taken to prevent the pest from entering the country. Should this pest arrive we must trust that the NBU have used this time well to prepare a strategy to combat it and we, as beekeepers, all get on board to implement their advice as to the best practice to control it.

Most diseases of the honeybee, except varroa, are quite rare, but when they occur, their effect upon the colony can be devastating. The best defence the beekeeper can have is knowledge. The National Bee Unit publishes a selection of pamphlets about disease and these are excellent. Even better they are free and can be obtained by simply contacting the National Bee Unit in York or through your local association. In return, it is right that we beekeepers should register on the NBU database 'beebase', a simple act that greatly assists the work of the bee inspectors.

Chapter 6
The Hive and other Equipment

A hive is a manmade container in which honeybees are kept. There is evidence of hives having been used for thousands of years. Columella, the foremost Roman writer on agriculture and beekeeping, describes hives being made from cork bark, basket work and pottery, the latter not being recommended. In this country the traditional hive was the straw skep, into which swarms were introduced as they were taken. In the autumn some of the skeps would be cleared of bees by killing them using sulphur fumes, and in this way the wax and honey was obtained. In the fifteenth and sixteenth centuries some individuals in England and other European countries experimented making hives out of wood, but these were expensive to manufacture. As the honeycombs were still attached to the sides and roofs of the wooden containers, as they were within a skep, this still didn't solve the problem of how to obtain the honey without killing all or some of the bees.

The Bee space

The major breakthrough occurred in 1852 and was the inspiration of the Rev L L Langstroth, a preacher in Pennsylvania in the USA. As the name suggests the Rev Langstroth's family originated from the Upper Wharfedale in Yorkshire. He realised that if a gap of between 6mm and 9mm is left within a hive the bees will neither fill the space with propolis nor build comb within it. This vital insight enabled hives to be built with movable frames, provided this proscribed gap, known as the bee space, was maintained at the sides, at the top and at the bottom of the frames. The bees leave the bee space intact and so the frames can be readily removed from and replaced into the hive. Within the frames the bees can be induced to build their comb, either by using foundation or a starter strip attached in the frame. This fundamental insight revolutionised the way that honeybees could be managed.

Comb

Hive body (super)

Beespace 6mm – 9mm

Frame runner

Frame

Hive body (Brood box)

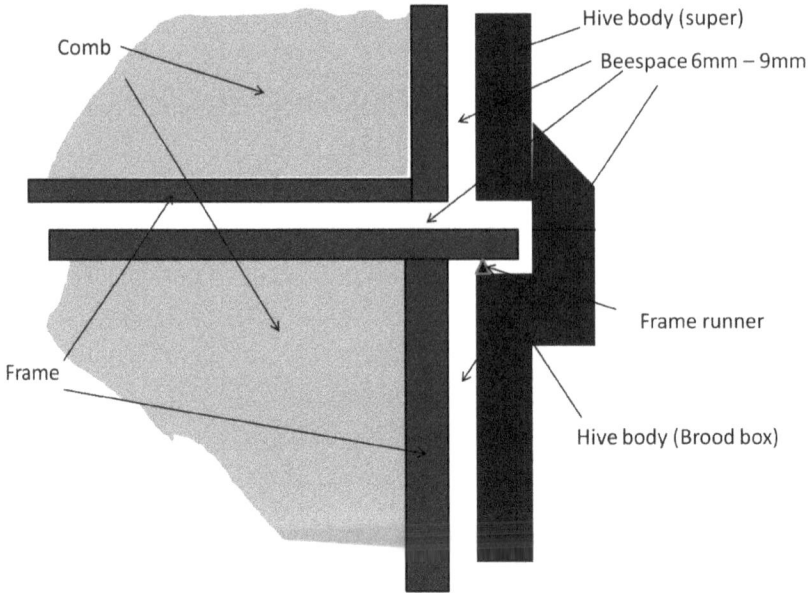

Details of Bee Space inside National Hive

Foundation is a sheet of beeswax with a pattern of tessellated regular hexagons, with the same dimensions as natural honeycomb, indented into the wax surface on both sides. Of course, this was not immediately available when Langstroth made his breakthrough but the implications of his discovery and the opportunities it presented to beekeepers led to a surge of invention and so it wasn't many years before foundation was being commercially manufactured. Foundation is not absolutely necessary. A 20mm, or thereabouts, wide strip of comb cut from an existing comb attached within the frame at the top will equally induce the bees to build a full comb within the frame.

The first hive to be built and patented with moveable frames was the Langstroth and this employed boxes that could be stacked on top of each other. As well as the consideration of the bee space, account must be taken of the volume that is required inside the hive. Studies of feral colonies have concluded that swarms prefer cavities of between 20 and 40 litres. The other parameter that must be considered is the spacing between the frames so that when the comb is drawn out in the frame, there is a bee space left between adjacent combs. Langstroth used 35mm while the designers of the National used 38mm. I have tried to measure the spacing of feral comb and could only confirm that indeed the spacing of 35 – 38mm seems right. It could be that Langstroth was using the smaller Italian bee (*Apis mellifera ligustica*), while the designers of the national would have studied the slightly larger British Black Bee (*Apis mellifera mellifera*), and so their natural comb spacing could have differed.

Hive Design

Once all these considerations became generally well known in the latter half of the nineteenth century, then anyone could design a hive, and indeed almost anyone did. The history of beekeeping is littered with hive designs no longer used. Other design criteria were taken into account such as the standard sizes of timber available and weight. In areas with colder winters it was necessary to use thicker timber to provide more insulation. In some areas of the world, because of the climate and available flora, honeybees are more prolific than in others and so require larger hives. Hive design also depends on the type of beekeeping practiced. Migratory beekeeping demands rectangular hives that can be easily stacked together. In large commercial operations weight is not an issue as fork lift machinery can be used to lift the hives. For the hobbyist, weight most certainly is an issue, especially for the more elderly.

Remarkably the very first removable frame hive, the Langstroth, remains the most popular worldwide, particularly in the USA and Australia. The modified Dadant is widespread in French speaking areas. In the UK the Modified National hive is the most popular but is by no means used universally. Other hive designs used in the UK include The Smith (mainly in Scotland), The Commercial, The Jumbo National (14 x 12) and the WBC. The debate about the most suitable hive continues to demand much time and consumption of beer.

As mentioned above standard hives consist of a stack of boxes. The bee space can be either at the top or the bottom of the frames. Langstroth hives usually have top bee space, while most users of National hives have a bottom bee space. Some hive types can be built either way and so it is a matter of the beekeeper's personal choice whether top or bottom bee space is used. Once again this issue has led to much consumption of beer without any conclusion being reached. Proponents of top bee space believe that this enables hives to be reassembled without damaging bees, as the hive boxes can be slid over each other without trapping the bees. Proponents of the bottom bee space do not allow this to be the case and add that you cannot use slotted queen excluders with top bee space configuration. What is agreed is that you cannot mix top bee space and bottom bee space equipment, so once a decision is made you need to stick with it consistently in all your hives.

The Floor

At the base is a floor that provides an entrance to the hive. Floors used to be solid and often incorporated a landing board. More often now beekeepers are using mesh floors. These have four advantages

1   They prevent the accumulation of debris at the bottom of the hive.
2   Their use is a part of an integrated pest management strategy to control varroa.
3   As the main source of ventilation is through the floor mesh, smaller entrances can be used throughout the year, which enables the colony to more easily maintain the security of the hive and so reduce robbing.
4   They prevent the build up of damp within the hive during the winter.

When mesh floors were first introduced there was much debate about whether their use was appropriate in the winter, but in our temperate climate in the UK honeybees seem to winter well on mesh floors. Indeed the extra ventilation through the mesh prevents the build up of moisture within the hive, reducing the likelihood of fungal type diseases. The mesh allows debris to drop through to the ground below and so there are fewer dark corners to act as refuges for wax moths.

The Main body of the Hive

The lower boxes are the brood boxes in which the queen lays her eggs and the colony raises the brood. Generally a single brood box is sufficient, but a significant number of beekeepers use double brood or brood and a half. Above the brood boxes are supers, boxes in which the bees are encouraged to store honey. By separating the brood area from the honey storage area, the beekeeper can remove honey with the minimum of disruption to the remainder of the colony. The supers are normally shallower than the brood boxes, but have the same horizontal cross section. Supers that are full of honey are heavy. A national honey super which is full of honeycomb can weigh about 14kg (30lb). For a lady or an elderly gent this is a significant weight, especially in view of the way one needs to lift a super from the hive, putting stress on the back. Any number of supers can be added to a hive during the summer and it's not unusual to see a hive with four or more supers on top of the brood box.

The honeybees will naturally use the bottom part of the hive to produce brood and the upper area to store honey. However most beekeepers use a queen excluder, placed above the brood box, to ensure that the queen, and therefore the brood are confined to the brood box. This ensures that the honeycombs are not contaminated with brood and makes finding the queen more straightforward. Queen excluders are of two main types, slotted or wired. The slotted queen excluders consist of a sheet of zinc, galvanised steel or plastic with a pattern of slots cut into them. The width of the slots is such that workers can readily pass through them, but the queen with her broader body cannot. The slotted queen excluders are cheap and easily cleaned. This is an important issue as the bees do tend to put propolis within the slots. The metal ones tend to have sharp edges and some beekeepers feel they can damage the bees' wings. The edges of the latest plastic queen excluders are rounded and are an improvement. The wired queen excluders consist of parallel wires, again spaced so that workers can pass through but the queen cannot. They are set in a frame having a bee space on one side. They are more expensive and are much more difficult to clean. However the wires are round and therefore do not damage the bees. There are alternatives to using manufactured queen excluders. One method that I have seen used is placing a square sheet of plastic, possibly cut from a fertilizer bag, on top of the brood box, but leaving a 3 – 4 cm gap around the edge to allow the workers to pass to and from the supers. And some beekeepers believe that queen excluders are simply not necessary. It is important that queen excluders are removed during the winter. If a super of honey is left in place above a queen excluder the cluster will not move into it, as this would result in the queen being left behind, stranded behind the queen excluder. The workers will not abandon their queen and so the colony can starve, despite honey stores being available within the hive.

On the top of the hive is a crown board. This is usually manufactured so that it can also serve as a clearer board. The crown board is basically a rectangle or square of plywood the same size as the cross section of the hive. In the centre are often cut shapes for inserting gadgets for clearing bees, the most commonly used being the Porter bee escape. These can be covered with a square of mesh which allows some ventilation but the bees invariably propolise it. In comparison I've never seen them propolise the mesh in a mesh floor. If the hives have bottom bee space, it is necessary to have battens, 6 – 9mm thick, the same width as the wall of the hive, attached to the edge of the board, in order to create a bee space over the upper set of frames. On the other side, thicker battens (about 20mm) should be used, so that the board can be used when treating the hive for varroa. It is common practice to have the crown boards in which the plywood is replaced with glass or another transparent material. This enables the cluster to be viewed during the winter without disturbing the colony.

A hive, brood box, queen excluder and two supers.
Note stand as described in text and mesh floor with restricted entrance.

Finally, above the crown board, there is a roof, normally flat and covered with galvanised metal. The roof provides a space for insulation in the winter and traditionally it included a gauze covered ventilation hole. However, since the introduction of mesh floors I have concluded that top ventilation is not necessary. The bees appear not to like through ventilation and invariably will use propolis to block mesh put over holes in the crown board. In recent years I have incorporated some insulation in the roof space all year round. Even in midsummer in the UK the temperature of the brood nest is at least 15°C above the average ambient temperature of the air surrounding the hive, so at all times of the year there is a loss of heat out of the hive. On occasions, for instance when the bees are processing honey during a honey flow, the bees may need to remove excess heat and this can be easily done by fanning and circulating air through the mesh of the floor. In this way the bees have control over the temperature and humidity within the hive rather than being subject to an uncontrolled draught caused by the chimney effect through the hive.

Gabled roofs are available and though they look twee, they have the double disadvantage of being more expensive and less practical. When inspecting colonies most beekeepers like to use the upturned roof as somewhere to place the supers so that the bees in the supers are temporarily confined while the work is being carried out. This is not possible with gable roofs.

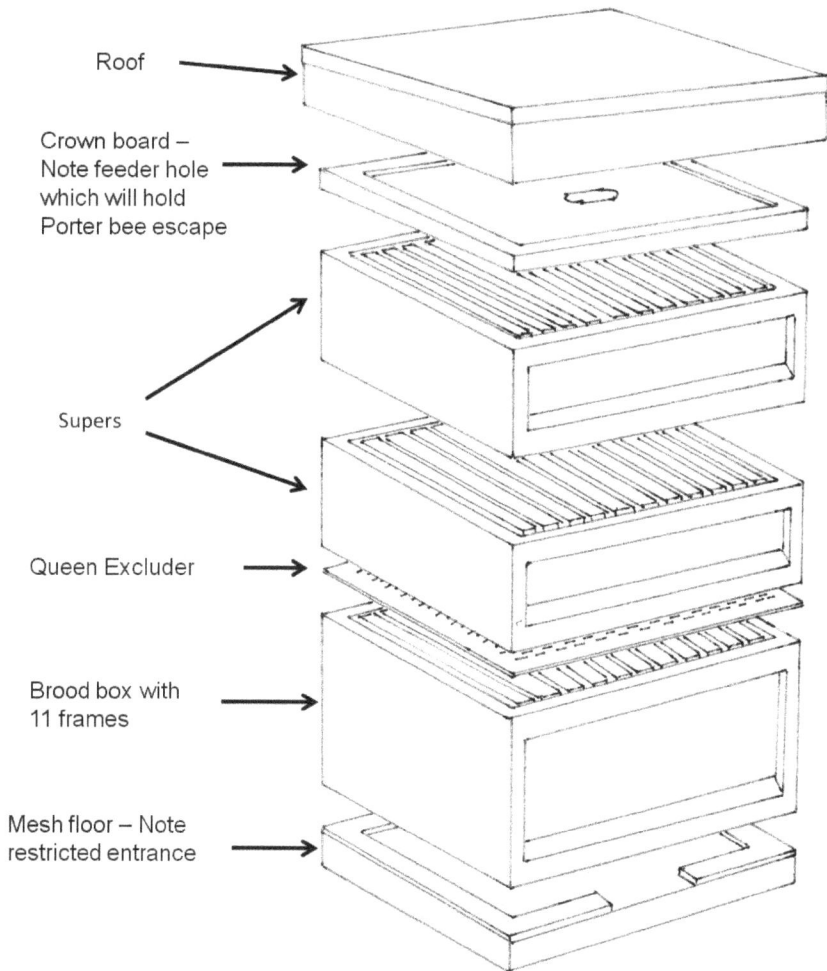

Roof

Crown board –
Note feeder hole
which will hold
Porter bee escape

Supers

Queen Excluder

Brood box with
11 frames

Mesh floor – Note
restricted entrance

Expanded View of National Hive

Choice of Hive

The above paragraphs are applicable to most single wall hives. All hives now in use have a long pedigree and therefore no matter what decision you make when deciding which hive type to use, it is never going to be absolutely wrong. Besides pure beekeeping reasons there are also considerations of cost and availability, especially of second hand equipment. The fact that a hive is commonly used in your area is a strong argument in its favour. Widespread usage will lead to an ample supply of second hand equipment and the cost of new is also likely to be less.

When deciding which type of hive is best from the bees' point of view, the following are things that you need to consider. The brood area needs to be large enough to give sufficient space for a prolific queen to lay at the peak of the summer as well as to store pollen and a little honey. The performance of the queen must not be prejudiced due to lack of space. On the other hand, if the brood area is bigger than necessary, manipulating and inspecting the bees becomes more arduous than it need be, the brood box is more expensive to set up and maintain and honey will be stored in the brood area rather than in the supers, compromising the amount of honey that can be easily harvested, as many radial extractors cannot process deep frames. Another criterion that needs to be considered is the capacity of the brood area to store sufficient honey for the winter, normally thought to be about 18kg (40 pounds).

Now let's introduce some mathematics to add substance to the discussion.

This may not be to everyone's taste but these are the calculations I did many years ago to justify the decision I made as to which type of hive I would use. I think it is interesting to understand why we have the hive designs that we do. Here I am trying to determine how much brood area is actually required. Most authorities suggest that a good queen will, on average, lay between 1500 and 2000 eggs per day at the mid summer peak. From observations of both my own colonies and colonies belonging to fellow beekeepers, I have concluded that the lower end of this range is more realistic, Fishermen tend to exaggerate the size of fish they catch, beekeepers the profligacy of their queens. There are occasions when there is a phenomenal amount of brood, but I believe this often arises when there are two queens laying in a colony during supersedure.

A worker requires 21 days from the egg being laid to the adult emerging, and therefore we can reasonably take 24 days as a cell turnabout period. So taking a lay rate of 1500 per day we get a total brood cell requirement of 36,000 cells. In practice the rate of egg laying for a given queen varies day by day depending on the weather and the quantity of forage that is available. Bearing in mind that an average of 31,500 (21 x 1500) brood cells will, assuming that an adult worker lives for 40 days, result in a colony size approaching 60,000 adult bees, it can be seen that this scenario represents a fairly strong colony.

Usage of cells in Brood Area

	Width mm	Depth mm	Area sq mm	No of cells per surface	No of cells on 20 surfaces	Percentage of available used for 36,000 brood.
National Brood Shallow	330	110	36300	1494	29880	
National Brood	330	190	62700	2580	51600	70%
Brood and a half			99000	4074	81480	44%
14 x 12 Brood	330	275	90750	3735	74700	48%
Lagsthroth deep	420	200	84000	3457	62226	58%

So how many cells are available. The area of a worker cell varies, between 5.1mm and 5.4mm in diameter. Taking an average of 5.3mm, this gives an area of 24.3 mm². In the table below I show the available brood area for some common hives and configurations. In a national hive 11 frames are normally used, but I'm assuming that the two outside surfaces are not utilised, and so this explains the 20 surfaces quoted in the table. Similarly Langstroth hives are normally set up with 10 frames, and so 18 surfaces are used in the calculations in the table. If the number of brood cells required is 36,000, it can be seen that a single national brood box would only be 70% utilised for worker brood. This leaves ample room for drone brood, pollen and some honey storage within the brood box. This is in accord with my own experience. I kept bees in the flat and exposed wheat lands in the Vale of York where a strong colony rarely consists of more than 20 full faces of brood comb. In other, more blessed areas, colonies may be larger. Nevertheless if 70% of the comb is to be available for raising brood, then it is necessary that the comb in all eleven frames is evenly and fully drawn and in good condition, and this in turn means that the comb needs to be fairly new, not excessively bound with pollen or old honey and undamaged. This will be returned to in the chapter on keeping healthy bees.

It can be seen that for a more prolific queen, say one capable of producing 2000 or more eggs per day, a single national brood box may not give sufficient brood space. In the table other configurations which are commonly used for brood are shown. Brood and a half references the practice of using a national brood box plus a shallow box for brood. This was generally recommended at one time and is still widely used. I do not favour it, mainly because it contravenes one of my fundamental beekeeping principles - 'keep things simple'. All my shallow boxes are set up for honey with castellated spacing giving 10 frames per box. The question of frame spacing is addressed in detail later in this chapter. If I were to use shallows for brood then it would be necessary to have some shallow boxes set up with 11 frames. Confusion would result. I also choose not to use 14 x 12 national brood. I know beekeepers who use them successfully and are passionate advocates for them. I believe they add expense, can be very heavy and the frames are too big to handle comfortably. For powerful colonies I choose to use double brood, which consists of two standard national brood boxes. By so doing I do not introduce another type of kit and the brood frames are standard BS deep throughout my apiaries. Using double brood gives a high degree of flexibility. By using dummy boards, I can choose to have 7 plus 7 frames (14 brood frames in total), 8 plus 8 (16 brood frames in total) or 9 plus 9 (18 brood frames in total), the latter being unlikely to be fully utilised by even the most prolific queen. To have two full brood boxes of frames is likely to lead to honey being stored in the brood area. By using dummy frames to control the amount of brood area available allows the beekeeper to extend or restrict the brood area as required. The double brood system also lends itself to methods of swarm prevention that are explained later.

Below are tabulated the more common single walled hives used throughout the world. For each I give their external and internal dimensions. It can be seen how the internal dimensions relate to the frame sizes in order to maintain the bee space. So, for example, for the Modified National , the BS frame, with a width of 356mm, fits into the 374mm internal width. The space at each side of the frame is therefore (374 − 356) / 2 = 9mm, which is the bee space as would be expected. The 11 frames must fit into the internal width of 424mm, which is 11 x 38mm + 6mm for the extra bee space at the side. Except that the spacing between the frames is allowed to be 35mm, a similar confirmation of the theory of hive design would be found looking at the Landstroth hive.

The WBC hive is not listed above as it is a double walled hive. It was designed by William Broughton Carr at the end of the nineteenth century. He was one of those big characters that dominated British beekeeping at that time. It's a pity we haven't such characters in these modern times. Being a double walled hive the WBC is double the cost and requires double the work to inspect the colony of bees within. But they remain popular with beekeepers who value the traditional authentic look. Bees seem to thrive in them. As they use British standard frames it is possible to run WBC hives and modified national together.

When I started beekeeping it was recommended that I use modified nationals which I have done for 25 years. I don't regret the decision. The only alternative I might consider if I started again is Langstroths.

DIY

Many beekeepers like to build their own hives. This is an excellent excuse to retire to the peace of one's shed during the winter months. The only things I wish to say is that I recommend using good materials and I do not favour the use of plywood on the external faces of the hive. Even good quality external grade plywood will begin to delaminate after a few years. By contrast hives made from deal and treated with preservative can be expected to last at least twenty years, longer in many cases. I own hives that I bought second hand twenty-five years ago and which were at least twenty years old at that time and they remain perfectly serviceable. The preferred wood used by hive manufacturers is western red cedar. It is long lasting and light weight. It is not readily available to the amateur DIY hive builder. In fact extreme care is required if you do use western red cedar as the dust produced is damaging to the lungs and it should not be used unless your workshop is equipped with effective dust extraction equipment and even then masks must be worn when using power tools. Hives must be built accurately to the millimetre. This is important when you understand the significance of the bee space, but too many beekeepers think it is a stricture that can be ignored. If you buy second hand items always go with a measuring tape to check the dimensions. Plans can be obtained from the BBKA website. Whereas making brood boxes, supers, floors and crown boards is quite straight forward, there is no point in making frames – they are cheap and require sophisticated machine tools to manufacture. If you are going to make equipment, it is definitely a winter activity. Once the spring is upon you there will be no time. Not only will you have the bees to attend to but there is also the garden. The winter months should be used to plan, and then procure or build any equipment that you think you are going to need. Beekeeping suppliers do occasionally hold sales during the winter. Spare equipment is absolutely essential, so don't just aim to have sufficient to contain your planned number of colonies.

Dimensions of most widely used types of hives

Type of Hive	External dimensions mm	Internal Dimension mm	volume (litre)	Frame size mm	No of frames	Area (approx) of each comb mm2	Top / Bottom space	Country of Usage
Modified National (deep)	460x460x225	374x424x225	35.7	356x216	11	62,700	bottom usually	UK
Modified National (shallow)	460x460x150	374x424x150	23.7	356x144	9,10,11	36,300	bottom usually	UK
Jumbo national 14x12	460x460x315	374x424x315	50	356x299	11	90,000	bottom usually	UK
Commercial	465x465x267		49	406x254	11	90,000	bottom	UK
Smith	464x416x225	378x419x225	35.6	357x216	11	62,700	top	scotland
Langstroth deep	508x413x243	468x368x243	41.6	448x232	10	86,900	top	US, australia, Worldwide
Langstroth shallow	508x413x146	468x368x146	28.2	448x135	10	46,000	top	US, australia, Worldwide
Dadant	508x470x298	468x425x298	59.2	448x290	11	111,000	top	French speaking

In the last decade, polystyrene hives are becoming more readily available. They are cheap, strong and light weight, all very positive qualities. They also provide better insulation for the bees during the winter. They are undoubtedly becoming more popular. Nevertheless I have a number of misgivings. Though the initial cost of purchase is less than wooden hives, I question whether the long term cost of ownership is less. A wooden brood box can be expected to last at least 30 years, but I doubt whether the life of a polystyrene hive will be anywhere near as long, as they are so vulnerable to damage from hive tools, rough handling, rodents, woodpeckers etc. They cannot be sterilised with a blow torch flame as you can a wooden hive. It is possible to sterilise them in other ways, but it takes longer. They are bulkier and so take up more space in your storage shed. The end of life disposal is more problematical. You cannot place your smoker on top of them as you can with a normal hive. I have not as yet tried a polystyrene full sized hive, but have several polystyrene nucleus hives, which are so cheap that they are not much more than disposable items. The better insulation allows small colonies to be overwintered successfully and I like the integrated feeders incorporated into the crown board. For my own part I like wood, owning wooden things and making things from wood. My father was the same and his father before him. Maybe it's genetic! I'm quite prepared to accept that beekeepers with a different genome can be content to use polystyrene hives.

The Frame

The standard frame consists of a top bar, two side bars, two bottom bars and a wedge. For those who wish to wire their own foundation in situ single piece bottom bars are available. Frames are usually purchased as individual parts and are assembled by the beekeeper, at the same time as the sheet of foundation is put in place. 18mm gimp pins are used to nail the frames together. If you are not happy using a hammer, then a rampin gadget can be used to insert the gimp pins.

There are a number of designs of frames, all of which come in the various sizes required to fit the standard hives. They differ in the width and shape of the bars. The normal British standard frame has a top bar and side bar widths of 22mm. The top bar can be obtained with a width of 27mm. Hoffman type frames have shoulders on the side bars which make them self spacing. BS frames used in modified national hives have a long lug which makes the frames relatively easy to handle, compared to the short lugs of the Langstroth frames. The frames rest on runners on which they can slide, or within castellations.

Frame Spacing

The failure to maintain the bee space between combs is the single most common mistake that new and not so new beekeepers make. It is an error that results in a confusion of comb in the hive and can take a lot of time to correct. Once again a decision is required. The new beekeeper needs to decide which method he wishes to use and then use that method throughout all hives. In the past, on a couple of occasions I've decided to alter my spacing method and it was a time consuming task converting all my equipment. There are three main methods by which the spacing between frames can be maintained.

Spacers on the lugs

Spacers can be slid onto the lugs of the frames. These may be plastic or metal, though I would not recommend the metal ones which are sharp and can cut your fingers and tear latex gloves. The spacers are 38mm or 50mm wide. The narrow spacers allow eleven frames to be inserted into a national brood box. When using plastic spacers it requires care to ensure that the edges of the spacers accurately abut against each other when reassembling the hive and the temptation to insert a frame without a spacer, simply guessing the frame spacing, must be resisted. When renewing foundation the spacers need to be removed and cleaned and this is not always easy. The bees will usually have propolised the spacers in place on the frame. The spacers do not rigidly hold the frames in place and the frames are able to swing back and forth if the hive is being moved. Nevertheless the method is cheap and easy, and is often recommended for new beekeepers.

The wider spacers are only used in supers, with the intention of inducing the bees to produce wider honey comb which reduces the work involved in extracting and enables more honey to be stored within a super. Unfortunately they cannot be used with foundation, as there is then sufficient space between the sheets of foundation to allow the bees to build wild comb, which they are sure to do. Once the comb is fully drawn on standard spacers, then these can be replaced with wider spacers. However in my opinion the effort involved in monitoring and altering the spacers is not worth the advantage of obtaining the wider comb.

Hoffman Spacing

Hoffman frames have side bars that have shoulders that give automatic spacing. The frame spacing is 35mm which makes them ideal for the Langstroth hive. When used in the national there is almost room for 12 frames, but it is a tight squeeze. A better approach is to still use eleven frames of comb plus a dummy frame at one end, which can be removed first when you start a hive manipulation, giving room for sliding the frames on the runners. Hoffman frames are more expensive and some bees will propolise the join between adjacent frames making them difficult to move without disturbing the equanimity of the colony. Hoffman frames do give some support when a hive is being moved. In my opinion they are not suitable for honey supers, as the space between the frames is too narrow, but they are a popular choice for spacing brood frames. Plastic converters can be bought to change normal frames into Hoffman type frames. Frames for Langstroth hives always come with Hoffman spacing, at 35mm centres. The short lugs of Langstroth frames prevent the use of spacers on the lugs.

Castellated Spacers

This uses galvanised strips of metal with castellations stamped along one edge that can hold the lugs of standard frames. The strips are tacked on to the inside wall of the hive replacing the frame runner, or slotted into a narrow slit cut into the top of the rebate for the frame lugs. The castellated spacers, when used on national hives, are designed to hold either 9, 10 or 11 frames. The 11 frame castellation gives a frame spacing of 38mm, the 10 frame castellation a frame spacing of 42mm and a 9 frame castellation a frame spacing of 46.5mm. Castellated spacing has a number of advantages.

1   The frames themselves, once removed from the hive, are unencumbered with spacers and so easier to clean and extract the honey.

2   The area of contact between the frame and the hive body is minimal and so the propolised joint between the frame and the body of the hive is minimised. As a result if you are inspecting your bees every week, there is no need to use the hive tool to crack the propolis seal.

3   The castellations hold the frame firmly and restrict the frames from swinging when a colony is moved.

4   The accurate spacing of the frames cannot be subverted by laziness or bad practice.

5   Once the castellated strips are attached to the hive, then cheap basic frames suffice.

Although castellated spacing is widely recommended for the honey supers, many experts do not recommend their use in brood boxes. The reason for this, they argue, is that when removing a frame it is necessary to initially move the frame upwards and parallel to the adjacent frame, running the risk of rolling the bees over each other and damaging workers or even the queen. Though it is not normal practice, I've used castellated spacing in brood boxes for well over a decade and believe that this risk is minimal to the extent of being nonexistent. The initial vertical movement is only about 9mm to bring the lugs above the top of the castellations at which point it is possible to separate the frames horizontally. I've used castellated spacing in the brood box for many years and I've never knowingly lost a queen in this way or even damaged workers and in my opinion the small risk is more than outweighed by the considerable advantages of castellated spacing listed above.

In the supers I use 10 frame castellated spacing throughout all my supers. This allows me to introduce frames of foundation into the supers without the bees being tempted to build wild comb between the frames, which would occur if I used 9 frame spacing. I believe there is a significant advantage in opting for a uniform standard that can be used throughout one's equipment. It makes life simpler, and allows more opportunity to consider life's unavoidable complications.

Hive Stand

The hives should be placed on a stand so that they are about 250mm (10in) above the ground. There are a number of reasons for this. In nature honeybees prefer to select cavities for their nests well above the ground. Experience indicates that bees do not thrive in damp positions and having the hives on a stand allows the hive floors to stay dry. And then there are the practical reasons. Having the hives positioned about 250mm above the ground makes the job of bending over the hives to manipulate the brood box less of a back breaking task, and eases the job of lifting heavy supers. The use of mesh floors to control varroa requires that there is a space below the floor so that the varroa can fall to the ground below without the possibility of their regaining access to the hive. The bees may very well benefit from having higher hive stands, but that would result in greater difficulties in lifting a third or fourth super on to a hive.

It is possible to buy metal stands that fold up. These are expensive but are ideal if you wish to move your bees. I have made up a number of flat frames made up from 100mm x 50mm carcase timber. The frames are 460mm wide and about 1200mm long. I use 100mm carriage screws to fabricate them. These frames are set up on two concrete blocks, one at each end, placed on their edge. When I make these frames I liberally treat them with wood preservative and allow them to dry before taking them to the apiary. These frames each hold two hives. Some care should be taken to ensure that the frames are accurately erected to be horizontal.

The layout of the apiary needs to be given some thought. The stands need to have sufficient space about them so that you can move around them easily. Positioning them in a circle makes sense or in a random pattern. Adding features such as bushes, preferably those that will produce nectar, or a bench again helps the returning foraging bees to locate their hive and reduces the possibility of the bees drifting into the wrong hive. Certainly placing the hives in a straight line should be avoided. At the start of the season I aim to have just one colony per stand. This gives room to make increase without compromise. There is without doubt a limit to the number of hives that can be maintained at a single site without compromising the performance of the individual colonies

Alighting Boards

An alighting board is a horizontal or sloping board below the entrance that bees can land on when they return to the colony, before walking into the entrance. I am unconvinced that they are an advantage to the bees, which are equally capable of landing on vertical surfaces as on horizontal surfaces. Others suggest that it is a real advantage to a tired bee returning with a heavy load of nectar. Feral colonies do not have alighting boards. Nevertheless they are useful for the beekeeper who wishes to spend time observing the bees coming and going. Alighting boards can be built into the construction of the hive floor. This has a major disadvantage when you wish to move the hives, possibly migrating to the heather, as the hives will no longer neatly pack together on the bed of the truck or van in which they are being transported. As I don't regard them as essential, those that I have are built as separate items which fit beneath my standard mesh floors.

The Eke

This is a useful item to have, and I aim to have one available for each hive. An eke is a rectangular frame, between 25mm and 50mm deep, with the same horizontal dimensions as the hive boxes. It is used to make space above or below the brood box or super. There are a number of circumstances when they can be used.

a) To give space above the brood box when applying thymol treatment on trays.

b) To make space when feeding with fondant during the winter.

c) To create space between the brood frames and the floor to protect the cluster from the winter winds.

d) To create space between a clearer board and supers which are being cleared.

e) To provide a surround and protection when using a feeder.

The Tool Box
The Hive Tool

The hive tool is absolutely essential. An old screwdriver or a paint scraper is not a suitable alternative. It needs to be painted orange or yellow so that when it is mislaid or falls in the grass then it is easily found again. The hive tool should have a 30mm scraper at one end (used for splitting hive boxes and scraping wax and propolis from hive parts) and a J attachment at the other end which can be used to loosen frames by levering, using the adjacent frame as a fulcrum.

The Smoker

A smoker is not absolutely necessary, but it is such an icon of beekeeping that almost all beekeepers own one, if only to take on demonstrations outside parliament. A smoker consists of a cylindrical firebox, a cone shaped nozzle at the top and a bellows attached that puffs air into the base of the firebox. The cone nozzle is on a hinge, so that the firebox can be opened. The fire boxes are usually manufactured out of copper or stainless steel. In my opinion it is more sensible to buy a larger rather than a smaller model. There is nothing more annoying than the smoker going out through lack of fuel half way through a hive manipulation. In the tool box there also needs to be smoker fuel. To light the smoker I use a plumber's gas blow torch. This is not a cheap item but having spent too much time sheltering from the wind under the wheel arch of the van trying to light the smoker, it is a luxury that I feel justified in enjoying. The fire needs to be established at the bottom before filling the firebox to the top. The best fuel I've found is wood chippings.

Bee Brush

The traditional tool for brushing bees is a goose wing. But soft hair brushes are available from beekeeping equipment suppliers. To brush bees it is essential to use a brush specifically designed for the purpose. It is not essential equipment but is required if you intend to use the shake and brush method of removing frames of honey. The brush is also necessary when treating a colony with icing sugar.

Other items

Pair of scissors

Queen marking pen

Queen cage

Indelible ink marking pen

At least one pair of hive straps

Porter bee escapes

Chapter 7

The Beekeeping Year

This chapter is the story of the beekeeping year. Sometimes things go well and sometimes they don't. A hundred and one things can conspire to reduce the best of plans to total chaos. The unpredictable effects of weather and family have to be accepted but it's not always so easy to accept one's own errors, whether of ignorance or omission. As is true of all beekeepers, how I keep bees is governed by where I keep my honeybees, (formerly in the Vale of York on the banks of the River Wharfe and now in the beautiful wild country of Herefordshire), the number of hives I choose to keep (between twelve and twenty) and the objectives that I set myself (keep healthy bees with a minimum of complications).

By general agreement the beekeeping year is thought to start at the beginning of October.

The Winter – October to February

This period starts once the feeding of colonies has been completed at the end of September and the hives have been prepared for the winter. The bees will continue to forage for another six weeks or so, well into November. Ivy is the main source of nectar at this time. Of course, the occasions when the bees can forage become fewer as the weather becomes colder and the days become shorter. However, the honey derived from ivy is a small but useful addition to winter stores. During this period the hives should not be opened and combs containing brood exposed to the cold air of the atmosphere. The cold air may kill the small amount of brood that is present and there is no point in checking that the colony is queen right. If the queen has been lost at this time, there is nothing that can be done to correct the situation. Even if the colony was able to produce a new queen, she would be unable to mate as there are no longer any drones. It is simply necessary to visit the apiary occasionally to check that the hive entrance is not blocked and that the hives have not been damaged in any way by livestock, by the wind, by falling branches, by floods, by woodpeckers or by vandalism.

The apiary in winter

During October and November the queen will continue to lay, though at a decreasing rate. As the temperature becomes lower the colony will spend more time in a cluster, which will gradually contract upon itself as the weather becomes colder. On still sunny days, in the middle of the day, it is not unusual to see the bees flying, even in December and January, but at this time the purpose is more likely to be defecation rather than gathering forage.

These quiet days of winter give an opportunity for the beekeeper to plan for the coming year and prepare the equipment that will be needed. Unused brood and super boxes can be checked and repaired when necessary. I have an ongoing schedule of treating the outside of the supers and brood boxes, floors and crown boards with insect friendly wood preservative and scorching the inside with a blow torch flame, effectively sterilizing them. Unused frames also should undergo maintenance. It is not good policy to store frames with old brood comb, whether deep or shallow, as they will attract wax moths. At the earliest opportunity I render down old brood comb in my steam wax extractor. The wax can be given in exchange for new foundation. The frames are effectively sterilised having spent an hour in steam at 100°C and simply need a scrape before being furnished with new foundation. Winter is also the time for making new equipment, possibly experimenting with new ideas and designs.

Super frames that have been used to store honey and which have sound comb can be reused the following year without putting in fresh foundation. These are unlikely to harbour disease pathogens. Stored comb can be subject to attack by wax moth, which can totally destroy a box of comb and the larger wax moth can damage the woodwork of the boxes. The wax moth can be killed by fumigation, but there are easier and simpler precautions that can be taken to lessen the risk. Simply by only storing comb that has been used for honey significantly reduces the risk of infestation with wax moth. Access to the comb by the moths can be denied. The boxes can be stacked on a board and then sealed at the top with another board and a liberal use of duct tape. Wax moths do not survive frost and so it is good practice to store the boxes in an open barn, on a veranda or anywhere where they are exposed to winter frosts and yet protected from the rain.

The winter is also a time for reading and study. There are libraries full of books about honeybees and beekeeping and many associations have a library of beekeeping books. They may not prove useful, but, besides, it is always a pleasure to read books by people that share your own passions and obsessions. BBKA run an examination system. It is divided into two main streams, practical and theory. The basic examination, designed to be taken after one or two years of experience, is a mix of theory and practical and is a prerequisite for all other examinations. Though examinations, especially the written theory, are not for everyone, they do concentrate the mind and force you, as a beekeeper, to reassess your methods, and separate fact from myth. And there is plenty of myth. The written theory examinations, which are organised into seven modules, take place in March and November and so the winter months are the ideal time to prepare for them. County and local associations sometimes set up tutorial groups to help prepare candidates or you can enrol on the correspondence course which is run and administered by BBKA.

The treatment of varroa is dealt with in detail in the chapter 5 'The Pests and Diseases of Honey Bees'. One of the treatments used as part of an IPM strategy is trickling oxalic solution on to the cluster. The time to give this treatment is when there is no brood, in other words in December. This is a job that, ideally, I like to do between the opening of Christmas presents and Christmas dinner, when I've picked the Brussels sprouts and it has been made clear that I'm better out of the way. Actually, the day should be chosen rather more carefully. The important thing is that there should be no wind, so that when the hive is opened the cooling of the cluster is minimised. It can be damaging to open the colony at this time of year, but an effective treatment of varroa is vital. For each colony this treatment can be completed in less than a minute. There is no requirement to remove frames and so the disturbance to the cluster is minimal.

At the end of September, each colony would have been checked to ensure that it had at least 18kg (40lb) of honey stores, which is normally considered sufficient to last until the spring. However, a large colony or a spell of mild autumn weather extending well into November could have resulted in more stores being used than would normally be expected. It may be surprising to suggest that bees use more stores during a mild spell. But a mild spell attracts the bees outside and the energy used flying can be significantly more than if the bees stayed in a cluster. And though it may be mild there are unlikely to be significant sources of nectar. Of course, if the weather becomes extremely cold then this too will increase the rate of consumption of winter stores. From December onwards the apiary should be visited every two or three weeks, and at those times the weight of the colony should be assessed. An experienced beekeeper will probably do this by hefting, which is done by gently lifting one edge of the hive from the stand so as to estimate its weight. The less experienced or more methodical beekeepers may use a spring balance as explained in the chapter 8 'Observing and Inspecting the bees'. It is important to be aware that the usage of winter stores is not linear with time. The rate at which the stores are consumed increases during March and the first half of April as the weather is becoming warmer. At this time the queen is increasing her rate of laying, the colony is raising brood and it is necessary for the flying bees to forage for pollen. But at this time there is still very little nectar available. A large proportion of winter losses are due to colonies starving during this latter part of the winter.

If it is discovered that a colony is short of stores then it needs to be fed. Maybe the fault is yours for not ensuring that the colony had sufficient stores at the beginning of the autumn, and maybe unusual conditions are responsible. Either way the bees must be fed or they will be lost. While the bees are in a cluster, that is during October, November, December, January and February, supplementary feeding should be with fondant. This can be obtained through beekeeping suppliers or more cheaply through wholesale catering suppliers. I cut the fondant into blocks and press the fondant into large margarine containers or the foil containers used for take-away Indian or Chinese meals. These are then placed, upside down, directly over the cluster or over the feed hole of the crown board. The height of the brood box can be extended with an eke to surround the fondant container.

The cold of a normal English winter does not pose any major issues to the wellbeing of the honeybee colony. Even the coldest English winter does not approach the severity of winters experienced in Eastern Europe where honeybees survive quite satisfactorily. As external temperatures fall the bees tighten the cluster within the hive and provided there are stores of honey they will survive. Losses of bees during the winter of 2009/10, which was one of the coldest for several decades, were less than in previous years.

From the beginning of January, the queen will begin laying, starting with just a small area, but gradually this will increase. As the early spring flowers appear, on warm days a small number of bees will begin to forage, and it's not unusual to find honeybees on snowdrops even when there is still snow on the ground. As the brood rearing increases the bees will take advantage of the pollen from the hazel catkins and willows, and then spring flowers such as crocuses.

Early Spring - March and April

The queen, if she has survived the winter months, will now be actively laying, though the number of adult bees in the colony will still be decreasing, as the old bees that were born during the previous autumn will be dying at a greater rate than new bees will be emerging. On sunny days, for the few hours around midday, the bees will be actively foraging. At the beginning of March the pickings are still sparse. In late March the blackthorn comes into flower and then a little later the first of the fruit trees.

Even in January a honeybee can forage on snowdrops

Some winter losses are to be expected. Historically a level of 10% has been accepted and there is no reason to go on a guilt trip if this happens. A session of self flagellation is appropriate if the bees have died from starvation but occasionally the queen will die from natural causes during the six winter months and when this happens the bees have no strategy to recover and there is no intervention that a beekeeper could have made to remedy the situation.

Managing for Oil Seed Rape

At one time beekeepers would be content to see their colonies slowly building up during these months, without any ambition of getting any significant immediate return in the way of honey. But, since the beginning of the 1990's, in the lowland agricultural areas of England, and particularly down the eastern side, in the second or third week of April, the countryside is transformed into a patchwork of brilliant yellow as fields of oil seed rape begin to flower. Though oil seed rape honey is disparaged by some, mainly by beekeepers from areas where it isn't grown, it is a favourite of many honey eaters, and can be processed to produce a beautiful soft set honey. There is also a demand for it from people who suffer from hay fever brought about by oil seed rape pollen. It is commonly believed that eating a local honey containing the pollens of local flowering plants can have a desensitising effect and so reduce any allergic reaction. Having said that, I'm always careful not to make any claims for my honey, which cannot be substantiated.

For the beekeeper it is an opportunity to get a crop of honey early in the season. When I first started beekeeping, a strong hive could produce up to 20 - 30kg of oil seed rape honey, possibly more, during the five or six weeks that the oil seed rape was in flower. Since then varieties have been introduced that are shorter in height and seem to produce much less nectar so the bumper crops of honey are less likely. The oil seed rape remains an excellent source of pollen. The weather at this time of year is often unsettled and cool, and if the daytime temperatures do not reach at least 15°C the bees will not forage with much enthusiasm. And weak colonies are unlikely to produce a great deal.

Managing your bees to take full advantage of oil seed rape has its challenges. Ideally, the foraging force that will take advantage of the crop needs to be in place by mid April. Bearing in mind that a worker emerges three weeks after the egg was laid and workers only become foragers from the age of twenty days or so, the foragers that will exploit the oil seed rape result from eggs laid six weeks before, that is at the end of February. In normal circumstances, at the end of February, the colony will still be depending upon winter stores collected during the previous summer and early autumn. To produce this brood the colony will also require pollen. In this last respect I am fortunate that in both areas where I have kept bees there are ample sources of pollen from willow and hazel. Pollen substitute can be purchased and is a necessity if there are no other significant early sources of pollen in the neighbourhood. Occasionally one reads in the beekeeping journals about recipes for making up your own pollen substitute but I have never been able to obtain the ingredients listed. I suspect the recipes are more theoretical than practical. An alternative is to collect pollen at times of the year when it is abundant, store it in a freezer and then feed it back to the bees in early spring. I collect pollen as a food supplement for myself, and occasionally have enough to feed back to the bees. I think a better policy is to ensure there are early flowering plants, such as crocuses, willow, snowdrops etc, near the apiary. From the end of February, I selectively feed sugar solution in order to stimulate brood rearing. Once feeding has started, it may be necessary to continue until there is natural forage available in sufficient quantities. There is no point in stimulating the queen to increase her laying at the beginning of March and then failing to ensure that there are sufficient stores to feed the expanded colony during a cold spell at the beginning of April. The beekeeper needs to be careful not to feed excessively, filling the super frames with honey made from sugar syrup and so contaminating the future honey harvest. The whole question of feeding is so important that I am devoting the entire chapter 11 to the subject.

The bees next to a field of oil seed rape

In theory honeybees can forage up to two or three miles from their hive, but in practice and for obvious reasons, they usually target sources of nectar that are less than a mile from the apiary. The energy expended in flying long distances can severely reduce the efficiency of their foraging. However, honeybees will generally find and exploit oil seed rape even if it is at the limit of the foraging range. The nectar has a high concentration of sugar, about 40%.

It is usually recommended that bees should not be moved less than two miles as the bees will recognise landmarks and return to the original site of the hive. This is not the case early in the season, when a move of one mile is normally tolerated, but once the oil seed rape is in flower the two mile rule certainly applies. And so, if the hives need to be moved closer to the crop, the move should be completed before the oil seed rape flowers. Under no circumstances position the hives in the crop. When the crop has finished flowering, it becomes an impenetrable tangled mass, and the position of the hives will only become evident again as they disappear into the mouth of a combine. Not a happy thought.

Generally, farmers are happy to have honeybees next to their oil seed rape crop. They like us to think that they wish to do their bit for the environment and honey bees and they are doing you a favour, but they are also well aware, as are beekeepers, that, although oil seed rape is primarily wind pollinated, honeybees can increase the yield of the crop by 5%. That seems a modest increase, but where you have a 40 hectare field this can result in an increase in yield worth several thousand pounds.

The flowering time of the oil seed rape varies. It is usually an autumn sown crop, drilled in late August or early September as soon as the previous cereal crop is harvested, and therefore is well established by the end of October. When we have a mild winter, the oil seed rape will continue to develop through the winter months and be ready to flower before the end of the second week of April. Following harsher winters flowering can be delayed to almost the end of April. The beekeeper needs, among his other skills, to be an expert in assessing when it is likely to flower. From the beekeepers' point of view, it is better late than early. A late flowering crop gives more time for the honeybee colonies to build sufficiently so that full advantage can be taken of it. At one time farmers also drilled oil seed rape in spring, in which case it flowered in June, but now this appears to be practised less commonly.

Before the oil seed rape flowers, I carry out a preliminary spring inspection. If I'm satisfied that there is a laying queen, I would content myself with putting on a queen excluder and a super. It is still too early in the season to go through the colony inspecting each brood frame. If I have left a super on over the winter, I would drive all the bees down into the brood box prior to putting on the queen excluder. If the queen has been laying in the super, the nurse bees will come back through the queen excluder so that they can continue to tend the brood in the super, and in three weeks or less the brood that was there will have emerged. The queen will continue to lay in the brood box.

The main ingredients of honey are two monosaccharides, glucose and fructose. Oil seed rape honey has a high concentration of glucose relative to fructose and this causes it to granulate quickly, so quickly in fact that it can granulate in the comb in a matter of weeks. Because oil seed rape produces a very distinct kind of honey, any remaining frames of honey from the winter stores should be replaced with fresh frames. The frames that have been removed, if you are confident they are not the result of autumn feeding, can be extracted, and the rest sealed and stored to be used for feeding later in the season. Because of this propensity to granulate, extracting the oil seed rape honey raises difficulties. There are two approaches that can be taken, either separately or in combination. The first is to remove full super frames on a weekly basis when the honey is still liquid and extract them straight away using a radial extractor. To wait until the honey is capped is often to wait too long and the honey will have granulated. Even though the comb is not capped the honey may still be ripe, that is, it has a water content of less than 20%. This can be easily tested by shaking the comb over the super box and observing whether drops of nectar fall from it. If they don't it, it is reasonable to conclude that the honey is OK. A more accurate scientific assessment can be obtained using a refractometer. The second approach is to allow the honey to granulate, cut the comb from the frame and then extract the honey by warming it until the honey returns to liquid form, and then separate the honey from the wax. All this is dealt with in greater detail in the chapter on processing honey. If the second approach is preferred it is better not to use wired foundation, so the comb can be cut out easily. If unwired combs are required, rather than use full sheets of foundation, it is possible to use starter strips of foundation. These are strips of unwired foundation about 2cm wide wedged into the top of the super frame. Having been given a start of the first centimetre or so the bees will generally complete the comb in the frame, but it must be said that occasionally the whole thing will go haywire, and comb will be built at an angle across the frames. As it is all to be cut out, it's not a great disaster when this does happen.

Potentially, the oil seed rape will continue to yield for five or six weeks. Even when the flowers of the main part of the field have set, small remnants on the headlands or beneath tall shaded hedgerows, which flowered later than the bulk of the field will extend the period of the yield and these areas will continue to be preferred by the bees to almost any other source of nectar. Once the oil seed rape has completed flowering, where possible, the bees need to be given fresh frames of comb which are clean of honey. Even small amounts of oil seed rape honey left on the combs can cause the new honey, which naturally would stay liquid for a long period, to granulate.

The oil seed crop gives an ideal opportunity to collect pollen. There is such an abundance of pollen available, that collecting it at this time has no effect on the ability of the colony to raise its expanding population of brood. The dynamics of the colony are such that more and more workers are diverted to pollen collection until the colony's pollen requirements are met. Of course, there is a cost. The number of workers available for collecting nectar is reduced slightly and this affects the amount of honey that will be produced. The pollen can be used as a supplement in the early spring, for your own consumption or as a product to sell. I've built a couple of special floors which incorporate a pollen mesh. The pollen needs to be removed from the pollen floor every two days and dried. In processing the pollen it is interesting to see that the pollen is not totally from the single oil seed rape source. Even though the colony may be positioned next to a field of oil seed rape the bees will go to considerable lengths to obtain a number of alternative pollens. Scientists now tell us that a mix of a variety of pollens is vital for the well being of the bees. The bees already seemed to be well aware of this.

During the period through April and May there will be other sources of nectar – plum, apple, pear, dandelion and hawthorn. The colonies will be growing apace. By the middle of April, in Yorkshire, it is time to do a full spring inspection, whether they are on the OSR or not. Choose a mild still day. The objectives of the inspection are

a) To ensure there is a laying queen

b) To assess the strength of the colony

c) To check for brood disease

d) To renew the brood comb

e) To exchange the brood box and floor

The diseases European Foul Brood and American Foul Brood (which have been described in detail earlier) are, thankfully, relatively rare but not so rare that we can be totally confident that our colonies will never succumb to them. At least twice a year it is essential to go through the brood frames to specifically check that there is no brood disease. Normally when inspecting the frames of bees you would not wish to shake the bees from the frames. But in this case this is what has to be done, regardless of the disturbance caused. This requires a sharp jerk with the frame held within the brood box. It is only if the brood frames are clear of bees that the brood can be inspected closely enough to detect a ,brood disease should it exist. You then must run your eyes over every part of the brood comb, sealed and unsealed looking for any signs of a departure from healthy brood. Both EFB and AFB are highly contagious and can quickly spread to all colonies in an apiary and beyond to other apiaries. In the spring time colonies can come under stress. They are growing quickly and their requirements can easily outstrip the availability of nectar and pollen. In these conditions of stress, disease can take hold.

As a part of a strategy to reduce the risk of disease, it is good policy to give the bees, each year, a fresh brood box that, during the winter, has been repaired, painted with preservative on the outside and sterilized with a flame on its interior surfaces, and at the same time renew the brood comb. During the previous year the brood comb will have been reused by the bees up to eight times. The inner surfaces of the cells will have become coated with propolis and faeces from the larvae, the inner dimensions of the cells will have been reduced, the bees will have damaged the comb reducing its effective area and if the colony has come into contact with disease, it will be in the comb that the pathogens are carried. There has always been a recommendation that there should be a rolling program of brood comb renewal but it is now becoming accepted practice to replace any set of brood comb that has been in use for a year or more. I use both the Bailey frame change and shook swarm methods. The procedures have similar aims but they have different outcomes and there are different situations in which they are appropriate. These are described in detail in Chapter 12 – 'Keeping Healthy Bees'.

Late Spring and Early Summer

This is the busiest time of the beekeeping year. The patience and tolerance of spouses is stretched to the limit. For the beekeeper it's exhausting but it's fun! All the preparation of the previous months comes to fruition as the bees pile in super after super of honey. There are weekly inspections to prevent swarming, telephone calls from the public about swarms that have got away (not yours of course) and possibly a bit of queen rearing. OK, so not everything goes to plan, but it's still fun.

May and June are the months of swarming. During this period, you need to put in place a regular inspection routine. If you do not clip your queens, I would suggest the inspections should be once a week. Strictly speaking it is possible for a colony to swarm even with this routine, as a queen cell can be raised to the point of being sealed in six days if the workers move an egg into a queen cup. But a weekly routine is more manageable for the average beekeeper who may be working or have other commitments.

The details of looking for queen cells and the procedures to adopt if one is found is dealt with in chapter 9 'Swarm control and Prevention'. The weekly inspection also needs to check that there is sufficient room in the supers for storing honey. If the bees are in the middle of a nectar flow then an additional super should be added beneath the existing supers. This is termed bottom supering. On the other hand, if you believe the flow is coming to an end, add the super above the existing supers. This is termed top supering. Supers are not just there to contain honey but give space to accommodate some of the 50 – 60 thousand worker bees in the colony. The supers also provide storage for nectar which can take more than twice the volume that is used by the honey that eventually is formed from the nectar.

The population of varroa in a colony needs to be monitored. Even if the colony was treated in the autumn and again in the middle of the winter, there will be a small number of the varroa mites that will have survived and which will be slowly but relentlessly propagating themselves. Every four to six weeks the population of varroa should be assessed by inserting the sampling board for a few days to collect the mite drop. Again the details of how this is done and the interpretation of the results are set out in the chapter 5 'Pests and Diseases of the Honeybees'. And as part of your IPM strategy the brood can be regularly treated with icing sugar shaken on to the brood frames.

The forage available during this period will vary depending on the topography, whether you are based in an urban or rural area, and local agricultural practices. In the area that I live the oil seed rape will continue to be a source of nectar until the last week of May. For a glorious fortnight during May the hawthorn will have come into flower and then the sycamore. The sycamore flowers are green, delicate and retiring, but on occasions they can be a major source of nectar. Walking beneath a sycamore tree at this time of year you can suddenly become aware that the tree is alive with bees, bumblebees as well as honeybees. The sycamore trees do not seem to all flower at the same time and so the time when they are yielding nectar can be spread over several weeks. Unfortunately, both hawthorn and sycamore do not seem to yield nectar with similar generosity every year.

A honeybee foraging on the cherry blossom

Sycamore blossom

In the apiary – Blackberries in Flower

Throughout the UK, at the start of June, the landscape suddenly becomes rather drab. It's still green, very green, but there are no reds, oranges and yellows. There are few flowers, except for the delicate fronds of wild chervil, cow parsley as it's known, lining the lanes and country paths. The greater part of the land area is given over to cultivating wheat and barley, a desert as far as the bees are concerned. The fields of oil seed rape, once the flowers drop, become a brown tangled wilderness. Most of the grassland is 'improved' and devoid of herbs and wild flowers, and what clover there is has not yet come into flower. In some years there can be a field of field beans within flying range of the apiary. But on the whole the bees need to rely on what can be collected from the village gardens and bits and pieces on the hedgerows.

This dearth, the June gap as it is known, doesn't last forever, but it can be a dangerous period. The colonies are now large and they need forage, both nectar and pollen, to provide for the adults and allow the raising of brood to continue at the same rate. Because the oil seed rape honey granulates it will have been necessary to remove the supers from the colonies so that the honey can be extracted and as a result the honey stores may be severely depleted. A close watch must be kept on the bees to ensure that they are able to still flourish so that when the good times do come, and as beekeepers we are programmed to believe that they will, then our bees are strong and numerous enough to take advantage of nature's bounty. It may be necessary to feed at this time. This is not something that you would expect to do every year, but it is something you should always be prepared to do. There are areas of the UK where this may never be necessary and beekeepers there will probably roll their eyes in despair at such a suggestion, but in the wheat lands of the east of England it frequently is a necessity, and when it is necessary it must be done. Bees that are stressed because they are on the point of starvation, will be bees that are vulnerable to disease. We all know that if we, as working men and women, become run down, it is then that we become ill. So it is with our bees.

Feeding may also be necessary at other times. In an English summer, at any time, there can be extended periods of inclement weather. We saw that during the summers of 2007 and 2008. The summer of 2007 was the first really poor summer for several years, and so beekeepers were not prepared or expecting to have to feed their bees. As a result, many colonies were poorly prepared for the winter that followed and this, along with other factors, contributed to the cause of the exceptionally high winter losses which followed. The following year, though the weather was no better, many beekeepers had learnt the lesson and were better prepared. They fed their bees early and losses were reduced. Poor weather does not only affect the time available for bees to forage but also affects the ability of the plants to produce nectar. If it is too dry the plants produce reduced amounts of nectar. If there is insufficient sunshine then the nectar produced has a lower percentage of sugar than would normally be expected. Frequent and sudden torrential thunder storms can wash the foraging bees on to the ground where they can be lost.

In a normal year, by the end of June all is well again. Patches of clover appear in paddocks and lawns, though sadly not in intensive grazing land. Farmers are being encouraged to sow clover mixes along the conservation strips at field edges and this provides a source of forage for honeybees, bumblebees and other nectar gathering insects. The lime trees that line the road into our old village flower and these can produce large amounts of honey. Then the brambles along the hedgerows and wood edges flower and by the end of July the Himalayan Balsam will start to flower along the river banks. It is with a little hypocrisy that I concur with those that deplore the encroachment of this invader along our river banks.

As we move into July the swarming season is coming to a close. It is no longer necessary to inspect every colony on such a regular basis and bees can be allowed to get on with enjoying what is left of the summer.

Joining colonies

At this time, it may be thought necessary to reduce the number of colonies by uniting two together. This must be carried out for valid reasons, such as simply wanting fewer bees, wanting to eliminate old or poor queens from your stock or to produce exceptionally strong colonies to take advantage of late summer forage such as heather. Before uniting a weak colony with a strong colony, the question needs to be asked as to why the weaker colony is weak in the first place. It needs to be checked that it is not being ravaged by disease such as varroa or nosema or that there are not drone laying workers. These conditions need to be eliminated before a colony can be used for uniting.

Each colony has its own distinct odour which is used by individual workers to recognise bees from its own colony and confront intruders. Without the beekeeper taking precautions, bees of different colonies will fight to the death if placed in the same box. The most reliable way of uniting is to use the newspaper method. The colonies to be united either need to be already positioned next to each other in the apiary or one or both brought in from a second apiary. This illustrates the importance of having a second apiary. When moving colonies, you either move them more than two miles or less than two feet. It is not possible to directly join colonies that are in the same apiary and more than a metre apart. Otherwise the flying bees will return to their original site where there is no longer a hive. One solution is to move both colonies to be joined to another apiary.

If the one or other of the hives has supers, a clearer board should be installed to clear the bees from the supers. The queen from one colony should be removed. The brood box of the colony with the queen is placed on the hive floor in its original position. Above this is placed the queen excluder, a sheet of newspaper from a broadsheet and then an eke. It can be useful to come to the apiary prepared with the paper already secured to the eke, especially if it's windy. Several pin holes should be punched through the paper. On top of the eke place the brood box of the second, queenless colony and then the supers. After a few days the majority of the newspaper will have been removed by the bees, shredded and cast out of the hive entrance, and the two colonies will have intermingled with few casualties. The final task is to reorganise the colony, in most cases reducing it to a single brood box, retaining all frames of brood and pollen in the colony and removing empty frames and surplus combs of honey which can be sealed and stored for use in the autumn.

There are alternative approaches, all of which are based on temporarily disguising or confusing the hive odours. For instance, both colonies can be liberally sprinkled with icing sugar and then shaken into the same brood box. By the time they have removed the icing sugar they cannot tell who is who.

Moving Colonies

There are a number of occasions when it is necessary to move colonies of bees.

1   To take advantage of forage crops, heather, oil seed rape, borage

2   To bring a colony from a quarantine apiary to the main apiary

3   To bring colonies together before uniting them

4   To make colonies available for education and demonstrations

5   To remove bad tempered colonies from sensitive positions

Again I remind you that you either move your bees more than two miles or less than two feet. Colonies are best moved in the early morning or after sunset, and warm weather should be avoided. For the small-time beekeeper, to prepare for transport, the colony should be reduced to no more than a brood box and a single super. Even this reduced hive can weigh over 40kg, and is more than a single person should be lifting and manoeuvring into place. Indeed, moving bees is not a job for a single person. It is never worth risking your back to move a few bees. As the entrance is to be sealed, provision must be made for ventilation. If you use mesh floors, then for short journeys on cool days, no other provision is necessary. If you are still using solid floors or if the bees are being moved a long way, travelling for several hours, then the crown board should be replaced with a screen. This preparation should be done the day before the move.

The entrance must be closed when the bees have stopped flying after the sun is set. Many beekeepers use strips of sponge which are pressed into the entrance and then secured with duct tape. With the roof removed, the boxes of the hive need to be strapped together, using two parallel straps. When being transported the hives should be stacked on to a flat bed, either in a van, truck or trailer. Straps or ropes should be used to ensure that the load is secure. The hives should be positioned so that the frames are at right angles to the direction of travel in order to reduce sway when cornering. When driving care should be taken to accelerate and brake carefully. If the journey is over an hour the bees should be sprayed through the mesh screen with water every so often. Once the colonies are unloaded and put in their new position the bees should be released. It is better to return a day later when they are settled to remove the ventilation screen and inspect the colony.

August and September

While the lanes around the village are clogged by trailers carrying loads of wheat back to the farms and we hear the rumble of combines working late into the evening, the honeybees are also intent on bringing in their harvest. By the beginning of August the Himalayan Balsam is well in flower, a ribbon of pink stretching for miles along the River Wharfe. The bees love it and as soon as the sun has risen, they are foraging with exuberance, pouring out of the hive as the first of the sun's rays strike the entrance. It is easy to see that the bees have been to the balsam as the dorsal part of their thorax is powdered with white pollen. The rosebay willow herb is also in abundance. There is an urgency that is palpable and indeed time is short, for both the bees and the beekeeper. Most IPM strategies to combat varroa will include a four week treatment that should be completed before the weather cools once the autumn equinox is past. And before the varroa treatments are applied the honey needs to be extracted. Working back this means that the honey that is to be extracted should be removed from the hives in the middle of August. This requires time to be set aside and preparations to be made. See Chapter 13, Honey and Wax.

After extracting the honey, the supers with the wet frames can be replaced on the hives. In a few days the bees will have cleaned the frames of honey and the supers can then be removed for winter storage. As soon as the honey is taken from the hives, it is time to treat for varroa. This is dealt with in detail in the Chapter 5 on 'Pests and Diseases of the Honeybee'. For the remainder of the season the honey that the bees store is theirs to enable them to survive the winter. Where I lived in Yorkshire there is six weeks of active foraging available once the honey harvest is removed from the hives, until the balsam finishes at the end of September or in the first week of October.

The colonies are rapidly contracting at this time. At the end of August the colonies will expel the drones, the workers pushing them out of the entrance and not allowing them to return. They soon perish. The colony will now be filling frames, which were formerly being used for brood, with honey stores. Eight brood frames of honey weigh about 18kg (40lb) and that is sufficient to survive the winter. Many beekeepers prefer to retain a super on the brood box over the winter, and for very strong colonies this could be necessary. Towards the end of September the queen excluder should be removed. This is because a cluster will not move through a queen excluder as this would entail leaving the queen behind, and so it would be possible for a colony to starve even though there were ample stores in a super above a queen excluder.

At the end of September the amount of stores needs to be assessed. A single brood box would need to weigh about 29kg (64lb) to ensure that the hive contains 18kg (40lb) of honey, sufficient for the colony to survive the winter. An experienced beekeeper would simply heft the hive, others could use a spring balance. If there are insufficient stores then the colony must be fed, and this feeding must be completed before the end of September. After that time the bees will find it increasingly difficult to process sugar solution into honey. To produce honey requires warmth and invertase. To synthesise the enzyme invertase the bees need the protein from pollen. As the autumn passes, both warmth and pollen become scarcer.

There are a number of other minor tasks to complete before one can be satisfied that bees are ready for the winter.

1. The hive entrance needs to be reduced. Later in the autumn when the cluster has formed mice can recognise that the corner of a hive can provide a warm nest in which to spend the winter. This can be prevented by restricting the size of the entrance to 8 or 9mm in height or placing a mouse guard over the entrance. A mouse guard is a strip of metal, about 30mm wide and 450mm long, stamped with a pattern of 9mm holes, which are sufficient to allow the passage of bees but not of the mice. Beekeepers who use mesh floors, like myself, may be employing a restricted entrance all year through.

2. Above the crown board additional insulation can be placed to restrict the loss of heat upwards, the same logic as householders installing loft insulation.

3. The integrity of the woodwork of the hive needs to be checked to ensure that it is sound.

4. If there are woodpeckers in the area, it may be necessary to protect the hives from attack. To the woodpecker a hive is a tree with a rich source of insect food within. They can be discouraged by wrapping chicken wire around the hive or pinning old cd's to the hive. But it is something that must be monitored throughout the winter.

5. Fences around the apiary should be checked and repaired to ensure that livestock cannot enter.

Time should be spent reviewing what went well and what didn't. It does no harm to write up a few notes. And so the beekeeping year comes to an end.

The plans for the beekeeping year were based on prolific queens, long hot sunny days in summer and supers full of honey. Then reality intervened – queens died in April, swarms were lost in May and there were floods in July.

When you see a beekeeper, despite all evidence to the contrary, you see someone who is fundamentally an optimist.

Chapter 8

Observing and Inspecting the Bees

Colony Inspection

It is reasonable to ask the question as to why it is that after several millennia when the vast majority of beekeepers functioned without carrying out regular inspections of the comb, it has now become the core activity of what we do as beekeepers. The simple answer to this question is that since 1853, when Rev L L Langstroth discovered the significance of the bee space and invented the moveable frame hive, it is now possible for beekeepers to carry out regular inspections. But just because something is possible does not mean that it is necessarily the right thing to do.

We need to be aware that every time we open a hive there is some collateral damage to the colony. The bees go to extraordinary efforts to maintain a homeostasis within the hive, controlling the temperature, humidity and the hive aroma, which is a complex mix of pheromones produced by the bees and aromas from the nectars. The retention of nest scent and heat is important for promoting the health of the bees. When the hive is opened there is the loss of nestduftwarmebindung. This is a German term for the odour and atmosphere that the bees create within the hive. When it is born in mind that bees use pheromones to communicate with each other within the hive, this disruption to the environment within the hive is equivalent to the internet being down for a while.

There is a variation of temperature within the hive, but in the centre of the brood nest the temperature is generally maintained at about 35°C and it is vital to the bees to maintain this temperature. Brood can be raised at lower temperatures but the workers produced at lower temperatures seem to lack some of the communication abilities that the workers normally exhibit. The high brood temperature seems to be essential for the full development of the nervous system and brain function.

Opening the hive for inspection is a contributory factor to colonies becoming bad tempered and tetchy. It can be observed how passive colonies in the early spring when the first inspection of the year is carried out, can gradually become increasingly bad tempered as we move through the season, even though they retain the same queen. Of course, there will be other factors in play causing this change in behaviour besides the disturbance caused by the inspections.

There are advantages in regular inspections. In the first instance, they allow the beekeeper to monitor any preparations for swarming, and by so doing be in a position to put in place procedures to manage the swarming instinct. This can reduce the nuisance caused to neighbours by swarms appearing where they are not wanted and reduces the loss of opportunity to harvest honey that happens when a swarm escapes. Secondly inspections provide the opportunity to check for disease and take timely action if it is found.

The preceding paragraphs have briefly rehearsed the debate between natural beekeepers and those using standard moveable frame hives. A few years ago I dabbled for a while with natural beekeeping, running two or three Warré hives in parallel to my nationals. But since moving to Herefordshire I have reverted to simply using my moveable frame hives. At no point was the debate in my head as to the relative merits of the two approaches decided one way or the other, but the nationals were easier to move, and I no longer have the strength and energy to run two types of hive. I do not regard the experiment with the Warré hives as a waste of time as I believe it gave me some insights into beekeeping that I have found valuable.

A beekeeper is often judged by the way he handles the bees, and indeed I have great admiration for my colleagues who I see handling their bees with such confidence and sureness of touch. Even after well over twenty years it is a joy to open up a hive and see the industry and artistry of the honeybees within, to marvel at the intricacy and delicacy of the comb and the myriads of interrelated tasks and activities that are underway.

To a large extent a healthy colony of honeybees will look after itself. If a beekeeper wishes to open up the hive, he should always have some purpose in mind. New beekeepers tend to want to open up their hive every other day. As a learning experience that is justifiable but as a long-term strategy for looking after honeybees it is counterproductive and is very disruptive to the colony's well being. The longer I've kept bees, the less I open my hives, but at the same time I spend more time slouching about the apiary, leaning on the fence, simply looking at the bees as they enter and leave the hive.

Hive Records

For many years I shared an apiary and my beekeeping with a very elderly lady. The deal was that I supplied the muscle and she supplied the thermos of coffee, club biscuits and endless stories of bees and beekeepers, most of whom were long dead. Beekeeping was a more relaxed business in those days; half an hour looking through the bees and then three quarters of an hour sitting beneath an apple tree at the other end of the orchard drinking coffee. Behind her wrinkled brow there was an encyclopaedic record of each of her hives. She seemed to have no need for written records and notes.

She was unusual and remarkable in many ways, not just as regards her memory. But this is not the way I can operate. If it's not written down for me it's lost and I've seen enough of the ways of other beekeepers to know that this is the normal state of affairs. Our memories are nowhere near as infallible as we like to think. And so, for many years now I have kept hive records. There seem to be four approaches to hive records, (1) not bother at all, (2) for the digital age beekeeper there are the computer or mobile phone apps, (3) there are cards that can be kept in the roof space of the hive or (4) there are records that are kept in a notebook or folder that is carried about with the beekeeper. If you are going to pretend to keep bees seriously and you do not have a photographic memory, record keeping is essential.

Some beekeepers like to keep record cards in the roof space of the hive. I have tried this in the past, but there were a number of snags with this. Cards got mislaid and ended up in the wrong hive or they were damaged by mice or damp. However, my main objection to this method is that it is quite inconvenient to access them before starting work on the hives, which to me is important, and they are not available to be studied at home in order to select the better queens. So, I have opted for the option (4), a folder that lives on the passenger seat of my van during the summer months. At the beginning of each season I print out sufficient blank sheets for the year, one for every hive that has come through the winter plus the same again, for each swarm or new queen that I expect to get.

There are countless examples of record cards floating about in books and on the internet. The one I use has evolved over time and will continue to evolve in the future. In the current format that I am using, there is header information including the year, a colony number, apiary name, queen number, overall rating. For each inspection I write in the date, weather conditions, weight, hive configuration, queen status, brood status, number of comb faces of brood, temper, honey harvested and, if required a note. Many of the entry boxes require just a single character code. The codes need to be simple and unambiguous. If giving a numbered rating of a characteristic such as temper, four options are all that are necessary. Trying to subdivide further results in inconsistency.

The weight

This is particularly important during the winter and early spring, to monitor stores and early build up. It can be quickly determined using a simple spring balance to lift one edge of the floor a few millimetres and doubling the value read off the spring balance. A screw eye can be put into each side of the floor framework to facilitate this.

Hive configuration

This documents the boxes used in a colony. ie. BXS indicates a brood box with one super, separated by a queen excluder, BXSSXB – a colony undergoing a Demaree.

Queen status

Seen – indicate colour (W,Y,R,G,B) or a tick if unmarked

NS – Not seen

V – Virgin

QC – queen cells

Brood status

E – Eggs seen

L – Larva seen

S – Sealed brood seen

Number of faces of comb containing brood, whether eggs, larva or sealed. I don't concern myself with adjusting for part faces. If the comb face includes some brood, I count it. This item is not about determining an absolute size of the colony, but is used primarily as a means of comparing the development of the colony from week to week.

Temper

Defensive behaviour

1 - Aggressive, following

2 – Manageable – React to manipulation. Smoke required

3 – Gentle - some smoke may be required

4 – Very gentle, no smoke required. Can be handled confidently without gloves

Calmness

1 – Run from the comb

2 – Clustering on the edges of the comb

3 – Moving on the comb

4 – Calm and static

Notes

In the notes I include details of any manipulations carried out such as shook swarm, signs of disease and alterations to the hive configuration such as adding a super.

The overall rating is important. It is a judgement on the quality of the colony and its queen and indicates how each colony may be used. This rating may be adjusted over the months. At the present I have three possible values

X – The queen should be culled

B – The queen can be used for breeding

S – The colony is considered as suitable for production, sale or distribution.

There are two main reasons to maintain records. The first is to help the beekeeper manage his or her bees. Before opening up a hive, the notes from the last inspection need to be read so the beekeeper knows what to expect, whether there should be a queen and with what colour she is marked, whether or not the colony had started swarm preparations or whether the colony is expanding or not. The beekeeper will also know what temperament to expect which may influence how the bees are handled. The beekeeper will know what action to take if the colony is preparing to swarm, whether to create several nucleus colonies or whether to replace queen cells with those from better rated colonies if these are available etc.

The second reason is that records are essential is if you are attempting a bee improvement programme. Any breeding programme that aims for an improvement must have at its heart a selection strategy. Even if your bee improvement programme is simply based on selective culling, you still need to have data which can be used to decide which queen is for the chop.

All pretty standard stuff. One of the problems I had for years was applying a colony number to a colony of bees. I had seen that some beekeepers used a number that was permanently affixed to the brood box. The trouble with this is that colonies can be moved to fresh brood boxes, either as part of a spring clean or as a result of a swarm control procedure. After some experimentation with different approaches I eventually decided to attach a small hook to each brood box and nucleus and made up a set a numbers, 1 to 40, painted on plastic or correx squares, each with a hole drilled through it so that it could be hung on the hook of the brood box. It is these numbered squares that designate the colony number. If a colony moves to a new brood box, the numbered square moves with it. This works well. The numbers are large enough to be seen from the 10 to 15 metres, so that by referring to the record folder it is possible to plan the programme of inspection that is required without leaving the van. It is simple, easy to do, and whereas I'm sure it is not a unique method of dealing with the colony numbering problem, I've never come across anyone else doing it this way.

Hive Record

Year
Hive No

Date	Weather Conditions	Weight	Config	Brood	Queen	Temper	Notes

Shown above is my record card, but, let me repeat, the most important thing is to maintain records, not the actual format of the records you keep.

Frequency of Inspection

My first inspection of the year will normally be towards the end of March. If the colony has overwintered with a super, it is at this time I would wish to replace the queen excluder that would have been removed in the previous autumn, first ensuring that the queen is definitely in the brood box. It is unlikely, though not impossible, that swarming will occur before the second half of April and so it is not until then that I would start regular weekly inspections. The reason for inspections being separated by a week is that this is the maximum interval if you wish to be reasonably certain that the colony will not have swarmed between inspections. Later in the season, once a colony has been subject to swarm control procedures, the frequency of inspections can be reduced. The time from having a sealed queen cell to a new queen having emerged, mated and begun to lay can be three weeks or more. There is very little point in disrupting the colony during this period, and the likelihood is that more harm is done than good.

The swarming season is usually completed by the beginning of July. Inspections are still required to check on the health of the colony, to check whether the colony is queen right and to determine the state of stores, so that in the latter part of the summer, inspecting the bees every three weeks is normally sufficient, and can cease altogether from the middle of September.

Preparation

Before starting an inspection you should have in your mind what you are trying to achieve. To formulate these aims you should first have read your hive notes from the previous inspection that was carried out the week before.

Here is a list of possible aims

1. Determine that the colony is queen right
2. Determine whether the colony is making swarm preparations
3. Determine whether the colony has sufficient stores
4. Determine whether the colony needs more space
5. Determine the disease status of the colony
6. Determine the strength of the colony
7. Determine the temper of the colony
8. To read the stage and development of the colony

When hive inspections are carried out, there are a number of risks that should be avoided or at least be minimised.

*   Manipulations should be conducted so as to minimise the risk of the queen being inadvertently moved above the queen excluder. If this happens it may take a few weeks before it is realised what has happened and a lot of work to remedy the situation.
*   Manipulations should not result in bees being harmed
*   The manipulation should be carried out in such a way as to maximise the chance of seeing the queen.
*   During an inspection the colony is losing foraging opportunity and the brood runs the risk of becoming chilled. An experienced beekeeper will therefore have learned to perform the inspection quickly, but without undue haste.

It is not always possible to choose your time, but inspections are more easily carried out in fine weather in the early afternoon, when many of the foragers are away from the hive. In such conditions the bees are so occupied with their own business they hardly notice the beekeeper's intrusion. The smaller number of bees in the hive makes it easier to find the queen and examine the brood. The activity of inspecting the hive becomes much more pleasurable.

Similarly, there are times to avoid. Thundery or windy weather will cause the bees to be much more aggressive than usual. By opening a colony when it is cold runs the risk of chilling the brood, causing it to die. However, a light drizzle on a warm summer day is not necessarily a problem. It is also best to avoid a time when there is a possibility of a virgin queen embarking on a mating flight. If the newly mated queen returns to a hive that is open, she may become confused and drift into a neighbouring hive.

The Smoker

It is sensible to light the smoker before opening the hive and well away from the hives. Do not wear your veil when lighting the smoker as there is a risk that, before the lid of the smoker is closed, the smoker could suddenly flare up and the material of the veil catch fire, causing serious injury. Having the smoker lit does not necessarily mean it must be used and the amount of smoke that is used is optional. But if the smoker is lit, it is available should it be required. It also makes sense to use a large smoker as this will continue to burn for a long period whether or not it is being used. The fuel I use now is wood chippings from the garden. A plumber's blow torch is ideal for lighting the smoker and no kindling is required. To cool the smoke a twist of green grass can be placed above the fuel.

There is no clear consensus as to how the smoker should be used, though there are plenty of beekeepers with rigid views on the subject. In my view smoking bees has two distinct and very different purposes.

1. To subdue them
2. To drive them from areas where they may be harmed during the manipulation of the colony.

Standard practice suggests that the bees can be subdued by smoking the entrance of the hive a minute or two before opening the hive.

The rationale behind this is that the bees will react as if their colony is threatened by fire. Their instinct in these circumstances is for the workers to gorge themselves with honey, filling their honey stomachs. The bees become less aggressive, possibly because they are unable to arch their abdomen sufficiently to sting. This same effect is observed with newly emerged swarms where the bees have filled their honey stomachs prior to leaving their home hive, and are usually very docile. When the hive is opened after smoke has been puffed into the entrance, it is often observed that the bees have their heads in the cells of the honey comb, feeding themselves. The smoke can also be used in small amounts across the top of the frames if there are signs that the colony is becoming a little tetchy. In my view smoking does have an effect of subduing a colony, but at a cost. The colony aroma is temporarily destroyed and the normal activities of the colony are disrupted for a significant period after the inspection is completed while the workers reprocess the honey in their stomachs and return it to the honey comb. It should be said that some beekeepers even dispute that smoke does subdue the temper of the colony. They argue that the bees that sting are the guard bees, a small band of sister warriors, and they will stay at their posts come smoke or high water.

My view is that I use smoke as necessary. Before approaching a hive, my notes will tell me what reception I can expect. If the hive is known to be calm and non aggressive I would not smoke at the entrance before opening the hive, but if my notes tell me that the bees may be aggressive then I would be prepared to smoke the entrance before opening.

However, I do believe it is absolutely essential to be prepared to drive the bees from places where they may be injured, especially when the hive is being reassembled after an inspection. It is very bad practice to kill the bees during a manipulation. This is not pure altruism. Squashed bees will release alarm pheromones and this can result in raising the level of aggression in the colony. And later the squashed remains of the dead bees will be removed and should these bees have nosema then the housekeeper bees that carry out the undertaker duties will become infected.

You will need a hive tool. The type I use has a scraper / blunt blade at one end and a J-tool at the other. The blade can be used to separate hive components and the J-tool is used to gently lever the frames away from the hive runners or castellation bar, cracking the propolis seal.

Observations at the Hive Entrance

Much information can be garnered by observing the bees at the entrance to the hive. The simplest and most significant observation is seeing whether or not the colony is bringing in pollen loads. If pollen is being brought in, this invariably indicates that the bees have brood to rear or that they soon will have, and that indicates that there is probably a laying queen. In early spring this observation quickly settles the mind as to whether the colony has survived the winter with a viable queen. During a nectar flow in the summer one should expect to see bees pouring out of the hive in the early morning to start foraging. The number of bees which are leaving and returning give an indication of the strength of the hive. A healthy colony has a vitality about it that cannot be mistaken, as opposed to a colony that is struggling, either through disease or insipient starvation, and appears dispirited. A weak colony can become a victim of robbing and disease. Robbing can be spotted by observing the potential thieves zigzagging in front of the entrance, as they case the joint. The zigzagging is a way of testing the defences of the potential victim. Colonies that are preparing to swarm stop foraging and bees congregate at the entrance.

Weighing the Hive

Weighing is another method of obtaining information without opening the hive. It is possible to buy weighing platforms that can be put between the floor and the hive stand and which gives a continual reading of the weight of the colony. This is an expensive piece of kit and I have been tempted, but my instinct of keeping my wallet deep in my pocket has always prevailed. A much cheaper and flexible option is to use a spring balance, which costs the equivalent of a couple of jars of honey. These are calibrated in pounds and kilograms. I now work in kilograms. If you intend to use a spring balance in this way, all the hive floors should be modified by putting a screw eye into each side. The spring balance is then used to gently lift the hive, easing one side off the stand. The reading on the spring balance can then be doubled to give a good approximation of the weight of the hive. A more accurate value can be obtained by weighing both sides and adding the two values together. I always remove the roof as roofs tend to be heavy and vary in weight, and so the roof is excluded from the calculations.

If you are to interpret the weight of the hive it is necessary to have a number of data items tucked away in your head, ready for use in the apiary but they probably will be of no use in a pub quiz. Of course, they are not absolutely accurate, but they are sufficiently accurate for the purpose of assessing the state of the colony.

Weight of a brood box plus empty frames – 7kg

Weight of a super plus empty frames – 5kg

Weight of bees – 7700 bees weigh 1kg, and therefore 40,000 bees weigh 5kg

Weight of honey in a full super – 12kg

Weight of wax in a brood box – 1.5kg

Weight of floor plus crown board – 1kg

Weighing is particularly useful during the spring build up. During March the weight will be continuing to decrease and by frequently weighing the hive it is possible to monitor whether spring feeding is necessary. From mid April onwards the weight should start to increase, and then more rapidly once the OSR comes into flower. Later on, the weighing technique can be used to assess

a.   Whether additional supers are required.

b.   Whether the hives have sufficient stores during any dearth during the summer.

c.   Whether a colony has sufficient stores for the winter.

Though most beekeepers are numerate I will go through a couple of examples of the mental arithmetic required.

1) to ensure that a colony is heavy enough for the Winter. Consider a colony in a single brood box. To survive the Winter the weight should consist of

Floor and crown board	1kg
Brood box and frames	7kg
Wax in brood frames	1.5kg
Bees (15000)	2kg
Honey	18kg
Total	29.5kg

i.e. between 14 and 15kg when hefting a single side.

2) to ensure that a colony has sufficient stores to survive a dearth in June. Consider a colony with a brood box and one super. Its minimum weight should be

Floor and crown board	1kg
Brood box and frames	7kg
Super and frames	5kg
Wax in brood frames	1.5kg
Wax in super	1kg
Bees (40000)	5kg
Brood (20000)	3kg
Honey (sufficient to ensure survival for a week)	5kg
Total	28.5kg

– that is between 13 and 14kg when hefting a single side

Opening the Hive

You are now almost ready to open the hive. First there are some general points to be made. When inspecting or observing the bees try to avoid standing in front of the entrance. You become an irritant to the guard bees and the returning foragers can become disoriented, possibly causing drifting to other hives or you end up with a cluster of bees on the back of your bee suit.

The reason that experienced beekeepers do not get stung is generally because they handle the bees better. They haven't developed bee proof skin or some magical bee repellent property on their hands. Movements about the apiary need to be calm and unhurried. When you are handling the frames take care not to knock or drop anything and avoid sudden movements of any kind. When you replace frames and reassemble the hive take every precaution to avoid damaging or squashing the bees.

There is usually no reason to look through the supers in detail. It is sufficient to be aware of their weight and whether there is still space within them. The centre of interest is the brood box. In most cases the bees will have firmly stuck together the floor, boxes and crown board with propolis

Before removing brood boxes or supers from the hive the propolis seal that the bees will have created between the hive parts needs to be cracked. The blade of the hive tool can be used for this. The blade of the hive tool needs to be inserted at least 20mm into the join before using it as a lever to break the seal, so that the corners of the hive are not damaged. Western red cedar, commonly used to make hives and which is ideal for this purpose in many ways, does tend to be rather brittle. The super needs to be loosened all the way round before you try to lift it. Care also needs to be taken that frames from the box underneath are not attached in any way. Just before lifting the super away, tip the super on one edge and blow a little smoke into the joint to drive the bees away from the bottom of the super so none are squashed when the supers are placed down after being moved.

The supers should be placed at least two metres from the hive. They need to be far enough away so you cannot trip over them as you are working on the hive and there is no chance of the queen finding her way into the supers. It is common practice to place the supers diagonally on an upturned roof. However, I prefer to place the supers on what I call a manipulation board. This is a board the same dimensions as the cross section of the hive, i.e. 460mm x 460mm, with a rim 18mm x 18mm attached around its edge. If you are strong enough or working with another beekeeper it is advantageous to lift off all the supers together.

If you are able to remove all the supers together then there is no need to remove the crown board separately. This not only speeds up the process, but reduces the number of bees that are released to fly around.

If, because you are working alone, you needed to remove the supers individually, I would recommend using a second manipulation board to cover the stack of supers. I am hesitant about recommending the use of cover clothes as they can become disgustingly dirty and there is some risk that that they become vectors of disease. It is some years now since I used one. You could say that the same thing could be said of manipulation boards but then they can be easily and quickly sterilised with the blow torch that I take with me to the apiary.

The brood box is now exposed with the queen excluder covering it. Next, the queen excluder must be removed. A light waft of smoke can be used to move the bees away. If a wired excluder is being used it must be levered up gently all the way round with the hive tool to break the propolis seal, or if using a slotted steel or plastic type it may be peeled from the brood box, taking care not to allow it to ping off. As it is being lifted from the hive, the underneath side of the queen excluder should be carefully examined to ensure that the queen is not on it, replacing her back in the brood box if she is. Place the excluder so it leans against the front of the hive to one side of the hive entrance with the underside facing outwards. This gives the opportunity for the bees, which remain on the excluder, including the queen if she has been missed, to regain entry back into the hive through the entrance. What should be avoided is placing the queen excluder on or against the side of the stack of supers as by so doing you are creating a risk that the queen will find her way into the supers.

Smoking to remove bees from queen excluder

At this point I will say a word about the orientation of the frames relative to the hive entrance. The frames can be set up either parallel to the side of the hive containing the entrance or at right angles to it. The former is known as the warm way, the latter is known as the cold way. The cold way allows better air flow through the colony, the warm way protects the centre of the hive from draughts. From one feral colony that I recently examined in detail, I could draw no conclusion as to the natural preference of the bees as the entrance was in the corner. However, as the majority of beekeepers now use mesh floors and restricted entrances these considerations have less relevance. When inspecting the bees, it is easier to lift the frames if the frames are running at right angles to the way you are facing. So, if you wish to work from the side of the hive, as opposed to the back, the cold way can be more convenient. If you choose to work from the back and set the frames in the warm way, then get into the habit of starting the inspection with the frame furthest from the entrance. The guard bees naturally congregate near the entrance and they are the ones that are more inclined to be aggressive.

So it is best to leave the frame which runs parallel to and is adjacent to the entrance to the end.

Having exposed the frames of the brood box, the individual frames need to be lifted out and inspected. To reiterate – always ensure that you are clear on the purpose of the inspection. Start from a side frame and if using Hoffman spacing this side frame may be a dummy. If the colony is not at full strength it is a good idea to have a dummy frame at both sides, which can be removed first. The bees will invariably have glued each frame on to its runners with propolis, and unless you are using castellation spacing it will not be possible to lift the frame with your fingers. By gently inserting the J-end of your hive tool between the side bars of the frame and the side of the hive, you can then use the adjacent frame as a fulcrum to lever up the frame, breaking the propolis joint without any sudden movement of the frame. Do not try to lift the frame with the hive tool. Once both sides have been loosened the frame can be gently removed from the brood box, holding the frame by the lugs at each end with your fingers. Move the frame sideways a small distance and lift it clear, taking care not to roll or trap bees. If bees are tending to congregate on the lugs, use a little smoke to drive them away. While examining the frame, hold it over the brood box so that if the queen is present and is dislodged she will fall back into the hive. Then carry out the inspection you planned, looking at both sides of the comb.

Use of J-end of hive tool to loosen a frame

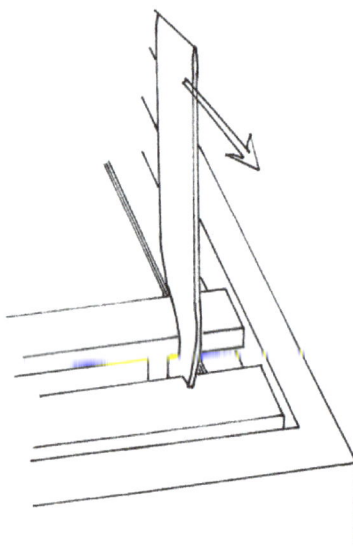

Do not replace the first frame that is inspected. It can be placed in a nucleus box, hung on brackets on the side of the brood box or placed horizontally on top of the frames still to be examined, moving it to the other side when you reach the half way point in the inspection. As a result, as you work through the frames of the brood chamber, there will always be a space left between the frames that have already been examined and those that have yet to be examined, allowing a wall of light into the body of the brood box. The space allows you room to remove the next frame without rolling the bees between adjacent combs. Also the wall of light discourages the queen moving from the frames waiting to be inspected to those that that have already been inspected. If the queen is on the frame next to the gap, she is likely to move around the side of the frame in order to be on the darker side. So when a frame is lifted out for examination, it is better to look at the dark side first. There is no certainty that this is where she will be, but she is more likely to be there on the dark side rather than the light side.

When turning the frame to view the opposite side, avoid going through a position where the frame is horizontal. With a little thought you will work out how this can be done. The frames can be quite heavy with honey, and even though the brood comb will normally be wired it is possible that the frame can collapse under its own weight, especially in warm weather. The bees would not be happy about it, and nor would you.

Healthy Brood

When you pick out a frame of brood to examine, you will automatically be checking that the brood is conforming to the paradigm of healthy brood. The brood nest is, roughly, an ellipsoid. The pattern you are observing as you view a frame is a cross section through the ellipsoid. At the centre of the nest the area of the cross section is greater than at the edges. A good queen will lay out the brood area with very few gaps, and, at the centre of the brood nest, the brood pattern will extend to the edges of the frame. Some queens do not like to lay above the wire reinforcement in the foundation, forming a large 'W' shape of empty cells in the brood pattern, but this need not be of concern. A brood pattern which contains too many spaces could be due to a failing queen or inbreeding.

There are three components to the brood, eggs, larvae and sealed. These form concentric areas, especially early in the season, though this pattern becomes less well defined as the season goes by. In theory it might be expected that eggs would occupy 1/7 of the brood area, larvae 2/7 of the brood area and sealed brood 4/7 of the brood area, but this assumes that the queen is laying at a constant rate. In practice the queen's egg production varies significantly depending upon the weather and availability of nectar.

The first priority when examining a frame is to try to spot eggs. These are white, about 1.5mm in length with the diameter of a fine thread of cotton. They are laid so that one end adheres to the base of the cell. Spotting eggs needs good eyesight, or at least good spectacles. It also helps if the wax of the comb is new, you are working in full daylight and are able to adjust the frame so that the sunlight is shining directly into the base of the cell. Seeing eggs gives the immediate reassurance that in the last three days the colony had a laying queen.

Healthy larvae are pearly white and lie in a 'C' shape at the bottom of the cell. The segmentation of the larvae should be quite clear. The cappings on sealed worker brood should be uniform, biscuit coloured, just slightly domed and look dry. They should not be damp, slimy or perforated.

Away from the centre of the brood there may be patches of drone brood, domed and somewhat larger than worker brood, and then further out, pollen. Above the brood area there will be some honey comb, often capped. Healthy brood is by far the norm, but the beekeeper needs to be prepared to spot when the exception occurs as this is when action needs to be taken.

Healthy larvae and freshly sealed brood
(Courtesy The Food and Environment Research Agency (FERA), Crown Copyright)

A frame of healthy sealed brood. Note the sealed honey comb in the top corners and
how the bees have the removed the comb along the bottom edge of the frame.
The comb is brown and looks ready for replacement even though it is
less than two years old.

Colony Checks

Each inspection will include an assessment of whether the colony is queen right and whether the colony is preparing to swarm. These topics are dealt with in detail in chapters 9 and 10. The inspection should also include disease checks, the details of which are covered in chapter 5.

Determine whether the Colony has Sufficient Stores

It only takes a few seconds but the inspection needs to include an assessment as to whether the colony has sufficient stores. It may take just a few seconds but there is quite a lot involved. During the summer months when the hives are being inspected every week, the beekeeper needs to be satisfied that there are sufficient stores to ensure that the colony can maintain itself until the next inspection. A large colony, during the summer, requires a large amount of nutrition, not just to provide for the metabolism of the 60,000 adult bees but also to nourish the growth of the 30,000 brood developing in the hive. The energy requirement is equivalent to about 5kg of honey per week. This is approximately what is contained in half of a full super, or two to three frames of honey in a national brood box.

Of course, much of the time during the summer the colony is actively foraging and during honey flows will be bringing into the hive at least that amount of energy nutrition in the form of nectar. The danger to the colony arises when there are prolonged periods of inclement weather or there is a dearth of nectar, as often happens in the June gap. So the judgement of whether the stores are adequate depends on not just the state of the colony, but the time of year and the prospects for the weather in the near future.

During the winter months the weekly honey consumption is much less but of course the colony is totally reliant on their stores. During the six or seven months of winter the colony uses between two and three kilograms of honey stores per month, a consumption that increases steeply during March and April when the colony starts to raise brood in large quantities. The assessment of stores during the winter can best be achieved by weighing.

Determine whether the Colony Needs More Space

Colonies of bees living in the wild have a fixed volume in the cavity where they had chosen to live. One of the major advantages of the moveable frame hive is that the capacity within the hive can be flexibly expanded and contracted according to the condition of the colony and the time of year. The colony needs space for two reasons, to expand the brood laying area and to store honey. A lack of space can lead to congestion within the hive and this can be a precursor of swarming.

During April, May and June while the colony is expanding, in the brood box there needs to be space at each side of the brood nest in the form of a frame of foundation or drawn comb. It is often the case that the brood nest is bounded by frames of honey of pollen. These are both potentially valuable resources for the colony and there are occasions when the colony will use or transfer these resources to other combs within the hive making space for new brood area. But often the beekeeper can help to make space for the expanding brood nest by moving combs.

More space can also be provided by adding supers to the hive. There are different strategies that can be used and all serve a similar purpose. During April and May once all existing boxes are more than two thirds covered with bees, it is an opportune time to add an additional super. And during this period when supers are added it is best that they are inserted directly above the brood box. There are a number of reasons for this. An expanding colony is at risk of congestion. By placing the super directly above the brood box it is able to mitigate congestion. During a honey flow, what the colony needs, even more than space for honey, is space for nectar. Nectar, (even if it is just a temporary situation) takes up two or three times the volume of the honey it will eventually become. It is better that the nectar is stored and processed close to the entrance of the hive and directly above the warmth of the brood nest.

If the super that is being added contains foundation rather than drawn comb, there is a further advantage in being placed directly above the brood nest. The workers will continue to move between the brood box and the existing supers and so the colony odour will be transferred to the new foundation. This and being in the warmest part of the colony will encourage the bees to draw out the comb.

Assessing the Colony

As you are inspecting the individual frames, the colony should be assessed. There are two main reasons for this, so that you can monitor how the colony is developing as the season progresses and secondly so you have comparative data that you can use to select which colony should be used for breeding. As you move from frame to frame

a. Count the number of sides of brood

b. Make a judgement as to the temper of the colony

c. Make a judgement as to the calmness of the bees on the comb.

These items need to be noted in your records. You could choose to record other characteristics such as uniformity of brood pattern or hygienic behaviour

Shutting down the Hive

Having completed the tasks that you have set yourself the hive should be reassembled, without panic or rush, but always bearing in mind that the colony should be returned to its proper state as quickly as possible. Care must be taken to ensure that bees are not squashed and a little smoke can be used to drive the bees from the tops and edges of the hive boxes. Like a surgeon counting up scalpels at the end of an operation, check that you still have your hive tool, and then write up your records straight away.

***

The apiary is my cathedral, a place to sit and contemplate, a place to stop and stare. With the eye mesmerised by the bees the mind is left free to wander. It is a place to get closer to your God, whatever form your God takes.

Find the Queen

Chapter 9

Swarm Control and Prevention

Swarm control and prevention are relatively recent introductions to the skills of a beekeeper. When I say recent, that is in the last 150 years, a mere moment in the history of beekeeping which stretches back thousands of years. Before the invention of movable frame beehives in the second half of the nineteenth century it was not easy to inspect the brood combs of bees within a colony on a regular basis. Beekeepers used straw skeps and accepted that bees would swarm. If they didn't want to lose the bees then the apiary would be visited regularly and a newly emerged swarm would be captured and placed into a fresh skep to form a new colony. At the end of the summer some of the surplus colonies would be destroyed so that the honey could be harvested. Swarming was an essential and necessary part of the management of bees. The occasional loss of a swarm would be accepted with a shrug of the shoulders.

Lost swarms do have a cost. The colony from which the swarm emerged will have lost at least half its worker bees and so is less likely to produce a surplus of honey. Not only that but it may have prejudiced its chance of surviving the coming winter without being generously fed by a beekeeper. Swarms are often regarded as a nuisance by the public. I like to regard swarms as part of the summer scene and the drama of nature. It is not always easy to make people understand that if left alone a swarm poses no significant danger, but we are obliged to try to preserve the good name of beekeeping and maintain good relations with our neighbours by preventing swarming from our own bees and collecting swarms when they are reported.

There is much written about swarm control and there are many books that are written to address this subject alone. I am making no attempt to be comprehensive in what follows, simply to set out the approach that I take. I do not clip the wings of my queens. This is not because of any false sentimentality, squeamishness or because it doesn't ever achieve the desired outcome. But I do try to have a uniform approach to the management of my bees throughout my hives and apiaries and beekeepers who do profess to clip wings never seem to have quite all their queens clipped. Clipping wings can reduce the frequency at which inspections are necessary but only by a few days. Instead of it being necessary to inspect every seven days it is possible to extend the time between inspections to about eleven days. This can be beneficial to the busy beekeeper, but a seven day inspection routine fits in with the other demands of life outside beekeeping, such as work and family, whereas an eleven day inspection routine doesn't and when the gap between inspections is stretched beyond eleven days, even with clipped queens, swarms can be lost. Clipping queens does not prevent the loss of casts, containing virgin queens. It can also result in the loss of the clipped queen, who attempts to follow a swarm, falls to the ground and is unable to return to the hive. And then the operation of clipping wings can lead to accidents, damaging perfectly good queens by removing the odd leg etc. Of course, it's all a matter of balance, weighing the benefits against the disadvantages. In my opinion the disadvantages outweigh the advantages. Others disagree.

There are two approaches to dealing with swarms, the reactive and proactive. Both are equally valid and beekeepers employ both in the management of their colonies. I certainly do. If this is not clear I'll first define these two terms

Reactive – Acting in response to a situation rather than creating or controlling it.

Proactive – Creating or controlling a situation rather than just responding.

Reactive management of swarming means that action is taken when it is seen that the bees are preparing to swarm by producing queen cells. To recap what has been written before, queen cells are capped 8-9 days after the egg was laid, and it is about this time that the swarm will issue from the hive. Therefore, during the swarming season, from the beginning of May to the middle of July, weekly inspections are required to check whether there are queen cells which contain eggs or larva. Often, without even visiting the apiary, it is possible to deduce that swarming is about to occur. Worker bees are seen sniffing about places where usually they are absent, in corners of sheds, around old chimneys or gaps in brickwork. These are the scouts starting to look for a new site for the swarm to build a colony.

The definitive indication that a colony is making preparations to swarm is the existence of queen cells containing an egg or larva. During the swarming season, May through to July, weekly inspections are necessary to look for these queen cells. If the queen cells are a prelude to swarming they will generally be found on the edge of the comb. In a feral colony, the queen cells are suspended from the catenary shaped edge at the bottom of the comb, but a comb held within a wooden frame does not have that free edge, either at the bottom or at the side. In these circumstances, the bees improvise, either by building queen cells on burr comb attached to the frame, by cutting away small sections of the comb at the side or bottom edges of the frame to make gaps in which the queen cells can be placed or building the queen cells parallel to the comb. In fact, the gaps in comb at the edges are a common occurrence. The bees don't always like the rectangular combs they are presented with. Maybe the gaps are made to provide corridors for movement or possibly for ventilation.

Queen cells are easily recognised. Unlike worker or drone cells they point downwards. They start as queen cups, looking like acorn cups if somewhat smaller, about 7mm across. The existence of queen cups cannot be taken as a sign that the colony intends to swarm. This is only the case if the edges of the cell are rimmed with fresh wax and contain an egg or larva. As the larva grows the walls of the cell are extended downwards and the cell will be about 25mm long when the larva is fully grown and the cell is ready to be capped. The walls of the cell are pitted like a peanut shell. Although, once it is spotted, queen cells are easily recognised for what they are, it is also easy to miss them. The mass of bees on a comb can easily conceal the cells. When examining frames the bees sometimes need to be gently moved aside to expose likely areas. It is not, however, advisable to shake the bees from each comb as this would risk damaging developing queens should queen cells exist on the comb, and is likely to distress the bees.

Some beekeepers recommend destroying all the queen cells to prevent swarming. I do not believe that this is generally a wise action, though it could have the desired outcome of delaying swarming if carried out by a beekeeper of sufficient experience. There are a number of dangers with this approach. If the queen cells are emergency or supersedure cells then destroying them could result in the colony becoming queenless. Swarm cells are a result of a colony being in a condition where it needs to swarm. Removing swarm queen cells does not take away the underlying swarming instinct of the colony, and the queen cells will be rebuilt. Furthermore, there is a risk that in going through the colony and destroying the queen cells, one will be missed. This one alone will be sufficient to cause the colony to swarm, and this will occur before the next inspection in a week's time.

Destroying queen cells is sometimes justified on the grounds that the beekeeper is not prepared to carry out a swarm control procedure because the required equipment is not at hand. This is a poor excuse. We know that swarming is more than probable and so throughout the swarming season the required equipment should be always there at hand in the apiary.

On occasions the correct action upon discovering queen cells is to do nothing. If the queen cells are emergency queen cells or supersedure queen cells then the colony is not preparing to swarm and you should note what you have found, reassemble the colony and wait for at least three or four weeks before revisiting the colony to check whether a new queen is now present and laying.

Even if you are confident that the queen cells are true swarm cells (over ten in number and positioned along the side, top and bottom of the brood combs) then you also need to satisfy yourself that the colony has not already swarmed. If the queen cells are capped then there is a high probability, but no guarantee, that a swarm has already emerged. On the other hand, in unsettled weather, the emerging of the swarm may be delayed, waiting for suitable weather conditions. The sure way of determining that the colony has not yet swarmed is to find the queen.

Now, having found a colony containing swarm cells and still containing a queen, then action must be taken to prevent a swarm emerging. The swarming impulse cannot be reversed and so the beekeeper must carry out a procedure that satisfies this swarming impulse by splitting the colony in a controlled way. The principle behind all methods of swarm control is to consider the colony as being made up of three parts, (1) the queen, (2) the foragers and (3) the brood plus the house bees. Swarm control procedures work by separating one of these from the other two.

Artificial Swarm or Pagden Method

This method separates (1) the queen and (2) the foragers from (3) the brood plus the house bees. The method requires a spare hive consisting of a floor, brood box complete with frames with new foundation, a crown board and a roof. At one time it was recommended to use drawn comb but now the demands of maintaining healthy bees require the use of new foundation. It is usual to have this equipment permanently available in the apiary during the summer months, set up to act as a bait hive to catch your neighbours' swarms.

Step 1. Remove any supers, placing them on a rebated board and covering with a second board or a cloth in order to seal them. Move the parent colony (A), see diagram, with the queen cells about three or four feet to the side, and possibly on the same hive stand and facing the same way as before.

Step 2. Place the new floor and brood box on to the original site, with the entrance oriented as the original colony. The flying and foraging bees will immediately start entering this new hive (B).

Step 3. Locate and trap the queen from the parent colony (A). Transfer a single comb of sealed brood, which must not contain eggs, young larva or any queen cells into the centre of the new colony (B). Place the queen on to this comb.

Step 4. Close down the new hive (B), replacing the crown board and roof. Later, in the evening, at dusk, feed the colony with a gallon of 2:1 sugar solution.

Step 5. Examine the parent colony (A) and destroy most of the queen cells leaving just one or two of the viable unsealed cells. If leaving two have them close to each other. Reassemble the hive, including replacing the supers.

Initial situation – brood box A contains queen cells

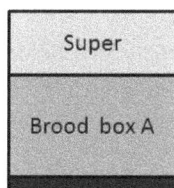

| Super |
| Brood box A |

Brood box A moved 1 metre to side. New brood box B with 10 frames of foundation placed on original site. Queen with frame of brood placed in B. Number of queen cells in A reduced to two. Super placed on A. Feeder placed on B.

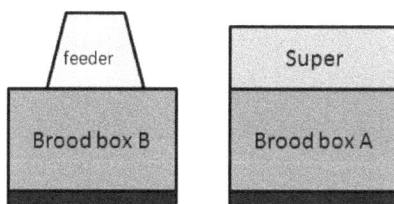

feeder

| Brood box B |

| Super |
| Brood box A |

After four days move A to far side of apiary. Remove feeder from B. Replace super on B

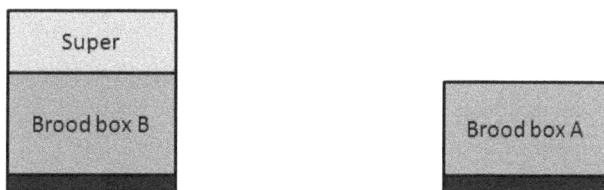

| Super |
| Brood box B |

| Brood box A |

Step 6. After four days, the feeder on the new hive (B) should have been emptied, much of the comb drawn out and the old queen should have resumed producing eggs. She will have with her all of the flying bees from the original colony, about half the original adult population of bees. The supers should then be transferred from the parent colony (A) to the new colony (B), and the parent colony (A) containing the queen cells and brood should then be moved across the apiary. This will result in a further influx of flying bees adopting the new colony (B) as their home. The colony B is unlikely to swarm again and so the frequency of inspections can be reduced. However, it is quite probable that the old queen in B will later be superseded. The parent colony (A) should be inspected again, removing any emergency queen cells and reducing the number of queen cells, now sealed, to just one.

In the parent colony (A) it will be between one and two weeks after the procedure is performed before the new queen will emerge and then when the conditions are suitable the virgin queen will be mated. In the meantime, the existing worker brood will continue to emerge, bolstering the population, and will have all emerged before the new queen starts to lay. Patience is required during this period. There is little point in opening and examining the colony until three or four weeks after the procedure is carried out, and it is possible that it could be even longer before brood appears. Indeed, it is necessary to avoid disturbing the colony when there is a chance that the virgin queen could be on a mating flight. A sure sign that all is well is when the workers resume bringing pollen into the colony. If the procedure is performed early in the season there is every chance that the colony (A) with its new queen can grow sufficiently to produce a small surplus of honey and certainly it should be sufficiently viable to survive the following winter.

As a variant on the procedure, it is possible to split the parent colony (A) into two or three nucleus hives, each one containing at least one queen cell. A nucleus is a small hive, designed to contained four, five or six frames. Each nucleus should be set up with two or three frames of brood with attached nurse bees, a queen cell, some additional nurse bees, a frame of stores, and a small number of frames of foundation. The size of the entrance to the nucleus should be minimised to no more than 15mm across. This procedure can be used to breed from queens that have demonstrated that they are able to produce strong productive colonies over one or two years and the bees are not aggressive. Making increase in this way will always come at a cost to the amount of honey which is produced. It will be necessary to continue feeding the nucleus colonies until they are well established.

The Nucleus Method of Swarm Control

As an alternative to the Pagden method there is the nucleus method. As before this is implemented when swarm cells are discovered and it can be seen that the colony has not swarmed as the old queen is still present. A nucleus colony is created containing the old queen, two frames of sealed brood with no queen cells and with nurse bees sourced from the parent colony, one or two frames of stores and the remaining space in the nucleus hive is completed with frames of foundation. In addition, one frame of nurse bees from the parent colony should be shaken into the nucleus. The remaining frames in the parent colony are closed up and the frames that were removed are replaced with frames of foundation. The parent hive remains on the original site, containing the forager bees, the majority of the brood and the majority of the house bees. Any sealed queen cells should be destroyed, leaving all the unsealed queen cells intact. One or two of the better of these unsealed cells can be selected and their position marked with drawing pins. By leaving the unsealed queen cells after the queen is removed, the colony is inhibited from raising further emergency queen cells. It will now be at least eight days before a virgin queen emerges from any of the remaining queen cells. The parent colony should be revisited after a week, fitting in with a normal inspection schedule, at which time all the queen cells, except one of the marked queen cells, which will now be sealed, can be removed. Also, at this time, as all the foragers that are so inclined will have returned to the parent colony, the nucleus hive can be fed. It may be necessary to continue to feed the nucleus intermittently until the colony has sufficient foragers of its own.

This procedure complies with the general principle set out above by separating (1) the queen from (2) the foragers and (3) the brood plus the house bees. It has the advantage of retaining a strong production colony, which, after a period of three or four weeks, will be led by a new queen. If the old queen is of proven quality you may wish to use her as a source of eggs for grafting. Otherwise she can be held in reserve until the new queen is established, and then culled and the colonies reunited.

Proactive Swarm Control

Proactive Swarm Control requires taking action before the colony wishes to swarm. We recall that swarming is triggered by the failure of the queen to produce sufficient of the queen substance pheromone to be distributed throughout the workers in the colony. This in turn may be the result of congestion, a very large colony or an aging queen.

The question of congestion is not always adequately considered. During May, June and July the supers are not just a space for storing honey but a space for the workers to rest and congregate. As a part of the weekly inspections, the beekeeper needs to be assessing whether there is sufficient room available. As supers are filled, then additional supers need to be added, placing the new supers below those that have been filled. The trigger I use to add an additional super, is when 70% of the frames in the existing supers either contain capped honey or are covered with bees.

After a queen exceeds the age of two years she will produce less queen substance. This will increase the chance of the swarming occurring. A policy of ensuring that queens are replaced after the age of two will reduce swarming and is regularly practised by bee farmers and beekeepers with large numbers of hives where it is impractical to examine all colonies each week. To manage this, it is essential that queens are marked and full records are kept of each colony. A policy of this kind usually requires a program of bee breeding and queen rearing.

The Snelgrove Method

There are several methods that can be used to pre-empt swarming, splitting the colony in a controlled way before the colony swarms in its own time. The method I have used is based upon the system set out by Snelgrove, an influential beekeeper from Somerset in the 1930's and 1940's. There are variants on his methods, but I choose to stick closely to the method set out in Snelgrove's book. When a colony has demonstrated that it has a prolific queen in the first weeks of May, the colony is converted to double brood, with eight brood frames in each box, confining the brood on each side with dummy frames. If, at subsequent inspections, it is found that the available space for brood is not sufficient then the dummy boards can be moved outwards allowing additional frames of foundation to be added.

About the first week of June the colony is split. The first stage is to reorganise the colony, putting all frames containing eggs or larvae in the upper brood box plus one frame of pollen and completed with combs of sealed brood and honey. The remaining frames of sealed brood and honey go in the lower box, and again the number of frames is made up with frames of foundation. The queen is placed in the lower box and confined there by a queen excluder positioned between the two brood boxes. The colony then needs to be left for several hours while the bees organise themselves after this drastic reorganisation of the frames.

Swarm Prevention using Snelgrove Board

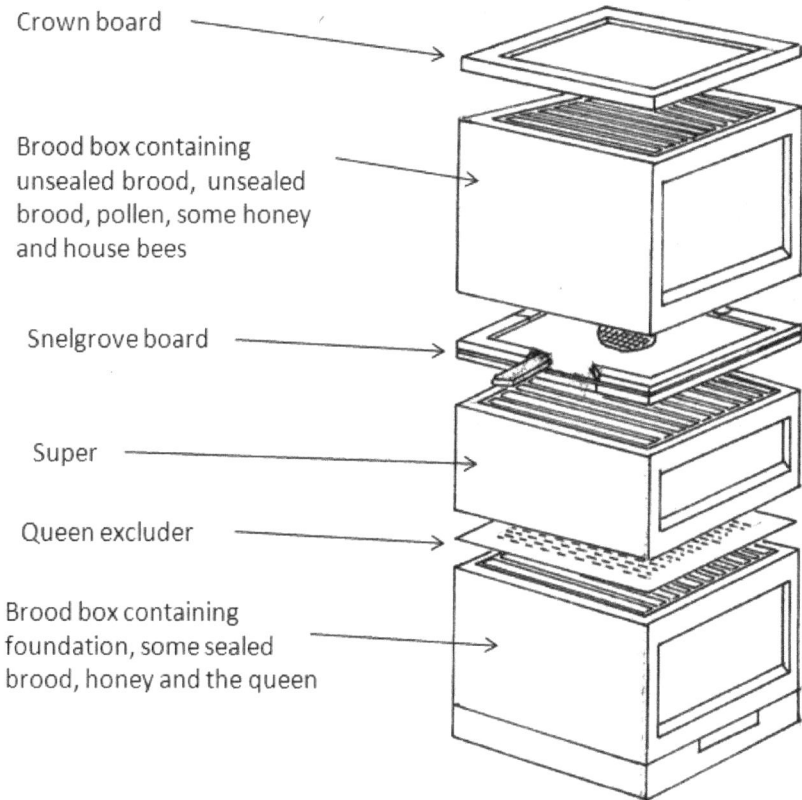

Crown board

Brood box containing unsealed brood, unsealed brood, pollen, some honey and house bees

Snelgrove board

Super

Queen excluder

Brood box containing foundation, some sealed brood, honey and the queen

Later in the day the colony is split. The upper box, containing the eggs and larva, and now most of the house bees is put to one side. The supers are placed directly upon the queen excluder over the bottom box which contains the queen. On top of the supers is placed a Snelgrove board, and above this the brood box containing the brood and eggs and finally a cover board. The situation we now have is a prolific queen in the bottom brood box, containing some sealed brood, a lot of foundation and all the foraging bees. In the brood box above the Snelgrove board are all the eggs and larvae, much of the sealed brood and most of the house bees, but no queen. The bees in the top box, isolated from the queen, will begin to produce queen cells. There will be large numbers of young bees emerging from the sealed brood in the lower brood box.

The Snelgrove board has a number of functions.

1. It isolates the bees in the top brood box from the queen.
2. It allows the two groups of bees to share the same colony aroma by having a mesh grill in the centre of the board.
3. It allows foraging bees to be bled from the top box to the bottom box so that the bottom part of the colony continues to be the honey producing area, even though it will be three weeks before significant numbers of new bees emerge in the bottom box. This is achieved by having six entrances, pairs on each of three sides of the board, built into the rebates of the board, three entrances opening into the space above the board and three opening into the space below the board. Each entrance has a simple hinged door that can be opened or closed when the board is in situ.

Snelgrove Board

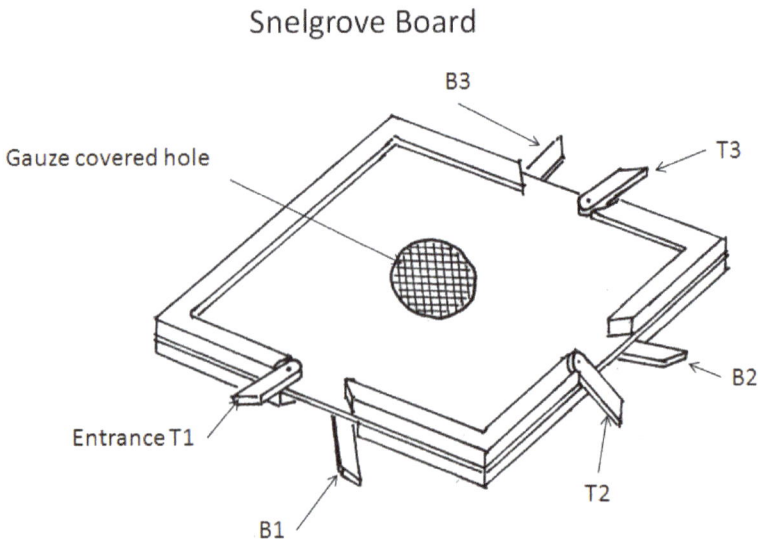

- 164 -

In order to explain the way the board is used, we label these entrances as T1, B1, T2, B2, T3 and B3, so that those prefixed with 'T' open into the top brood box and those prefixed with a 'B' open into the supers below the Snelgrove board, with T1 directly above B1. When the board is placed initially during the hive reorganisation the edge of the board with no entrances is positioned to be above the hive entrance of the bottom brood box. All the entrances of the board are closed except for T2, which becomes the entrance used by foragers from the top brood box.

After four days T2 is closed, T1 is opened and B2 is opened. In the four days since the manipulation was carried out significant numbers of new bees will have emerged in the top brood box, and many of the older house bees will have matured to become foragers. These foragers will continue to return to the entrance T2, but as this is closed, they will enter B2, and so become a part of the foraging force of the lower part of the colony. The younger foragers will use the entrance T1 and their numbers will continue to increase as the days go by.

After another four days, that is eight days after the manipulation, B2 and T1 are closed and B1 and T3 are opened. Again, this will result in the foraging bees from the top box that had been using entrance T1, joining the bottom colony through the entrance B1. By this time the queen cells that had been formed in the top box should have been sealed.

The final stage of this process is done after another four days, which is twelve days after the initial operation. B1 and T3 are closed and B3 and T2 are opened. Again, this will result in the foraging bees from the top box that had been using entrance T3, joining the bottom colony through the entrance B3. In the bottom box the old queen will be rapidly establishing brood on the new comb, the colony, being reinforced with the new foragers from the top box should, given the correct conditions, be storing honey. In the top box the queen cells should be sealed. The top box should be inspected and the number of queen cells reduced to one or two. Alternatively, frames can be removed from the top box, including frames containing queen cells, so that nucleus colonies can be created.

At a later stage entrance B3 can be closed. Virgin queens will emerge in a few days and in a further few days the dominant one will fly from entrance T2 on her mating flight, and eventually a new independent colony is established in the top box. Later in the season the new queen can be used to replace the old queen of the original colony.

This must seem very complicated, opening and closing doors in sequence. In reality it's not that difficult. The door opening and closing takes a few seconds for each colony. As it is the beekeeper that is in control of initiating the procedure, it is sensible that all the colonies in the apiary that are being managed in this way conform to the same timetable. This reduces the number of visits to the apiary.

When the Swarm has Left

There inevitably will be occasions when the swarm will have left before the beekeeper can take preventive measures. The colony will be left with all the brood, between ten and twenty mature queen cells, half the original number of adult bees and no queen. In the following three weeks the brood will have developed and then emerged as adults, returning the population to a figure approaching what it was before the swarm left. But by this time there will be no brood and so the population will then start to decrease. The first of the virgin queens will emerge about eight days after the prime swarm departed. If she is able to despatch the other virgin queens, then it will be at least another twelve days during which the queen matures, mates and is ready to start laying. Depending upon weather conditions it could take ten days longer. Once this happens then the colony will eventually start to recover.

An alternative outcome is that the virgin queens emerge successively and each heads a secondary swarm (or cast) that leaves the hive with half the remaining worker bees. This process very quickly destroys the viability of the original colony. This is extremely undesirable, but the beekeeper can intervene and, at the time when it is recognised that the swarm has emerged, by selecting the best queen cell and destroying the remainder, remove the option of the colony producing casts. It is best to select an unsealed queen cell rather than a sealed cell so a proper judgement can be made on the quality of the queen larva.

Chapter 10

Management of the Queen

The quality and condition of the queen is probably the single most significant factor that affects the performance of a honeybee colony. Her life expectancy is between two and four years and while she lives it is her genetics which will influence the industry of the colony, the temper of the colony, the immunity to disease, the ability to survive harsh conditions, the tendency to use excess propolis and numerous other characteristics of the colony. In the first two or three weeks of her life she will, in the course of one or two days embark on several mating flights during which she will mate with between ten and twenty drones. Then during the next day or so the sperm she received will migrate from the oviduct up the spermathecal duct into the spermatheca, a spherical organ towards the rear of the queen's abdomen. Then after several days she will start to lay eggs. Each egg that she lays into a worker cell will be fertilised by one of the few sperm that are released from the spermatheca as the egg is laid. A well mated queen will start her productive life with about seven million sperm in the spermatheca. When they are exhausted the queen's useful life is over. A drone is produced from an egg that has not been fertilised.

The Rate of Laying of a Queen

The maximum rate at which she lays is variously stated by different authorities as being between 1500 and 2000 eggs per day. In my experience the amount of brood I see in a good colony during the peak of the summer is consistent with a laying rate of about 1500 eggs per day, but I am sure there are queens in some areas of the UK and the world that exceed this rate. The rate of laying is not constant. At the winter solstice there is virtually no brood produced. From that point the egg production gradually increases during February, March, April and May. Even at the peak of the summer the rate at which the queen lays will fluctuate from day to day, reflecting the amount of pollen and nectar that is available to feed the brood that is being produced. During August, for short periods, it is not unusual for the queen to take a complete break from egg production and then resume when it is time to produce the winter bees. During the course of a full year a fecund queen can produce over 150,000 eggs. It would appear that it is the workers as a body that control the rate at which eggs are laid by the queen by controlling the amount of nutrition the queen receives.

The other role of the queen is to produce pheromones, which are collectively known as queen substance and which act as a cohesive force within the colony. When there is insufficient queen substance being produced then the colony will take steps to raise new queens and so proceed to split or swarm. As the queen ages her ability to produce queen substance decreases and so the likelihood of swarming increases.

Queen Right

Beekeepers use the term 'queen right' to indicate that, as regards the presence of a queen, the status of the colony is satisfactory and that no action is required. In simplistic terms this would mean that the colony has a laying queen.

Finding the Queen

The best way of being sure that the colony is queen right is by spotting the queen and observing that there are eggs in worker cells. This is not always easy. Even a marked queen can be missed during inspections by experienced beekeepers. However, the experienced beekeeper will either persevere until the queen is found or sufficient alternative evidence is observed to enable a confident judgement to be made that either a queen does or does not exist. Finding a queen is a beekeeping skill requiring good eyesight and experience. And the truth is that many new beekeepers taking up the pastime later in life have neither the experience nor good eyesight. Often taking up beekeeping encourages a visit to the optician, and it's very satisfying to discover how much a good pair of glasses can help.

In looking for the queen you are looking for one insect in tens of thousands. She is larger than a worker, but only by about 30%. Most of this additional size is in the abdomen, and as a result her wings don't extend so far down her abdomen in comparison with a worker. Also, the queen seems to have longer legs than a worker and so she walks across the comb in a different way. Sometimes I think it is marvellous that we find an unmarked queen at all. But we do. Some queens stand out while others do their utmost to remain undiscovered. One problem is that some queens, especially younger ones, move quickly on the comb and tend to run around the edge of the frame on to the darker side, and so the first thing to do, after picking up the frame, is to scan around the edge and then spiral in towards the centre. You basically need to think 'Queen', and allow your subconscious to take over your eyes. Once you get used to looking for her it's amazing how successful you can be. When I lift out a comb, I always examine the face that was up against the adjoining frame first, as this side would have been in the dark and there is a higher probability that the queen will be there rather than on the side that has been exposed to the light for a few minutes. In general, you would expect the queen to be in the centre of the brood nest but this is by no means always the case and so each frame needs to inspected with care. And of course, she always has the option of moving on to the floor or the sides of the inside of the brood box and this really can make things difficult. While smoking may make the bees less aggressive it also may drive the queen away from the centre of the brood nest and make her more difficult to spot.

Marking the Queen

A marked queen is much easier to find and it is generally accepted practice to mark the queen, with a spot of paint on the top of her thorax. Most beekeepers conform to the accepted standard for the colour of the spot, the colour indicating the year that the queen was produced.

White   Year Ending 1 or 6

Yellow   Year Ending 2 or 7

Red   Year Ending 3 or 8

Green	Year Ending 4 or 9
Blue	Year Ending 5 or 0

It can be seen that the marking of the queen serves two purposes. It aids the finding of the queen and it is a record of how old she is.

Marking a queen is a necessary skill for new beekeepers. The operation cannot be carried out wearing heavy leather gloves and so here is another argument for using disposable nitrile gloves. If you are right handed the queen is picked up with the right hand and transferred to the left so that she is held by the thorax, dorsal side upwards, securely between the thumb and first finger. In this position it is easy to dab a spot of colour on to the top of the thorax. While she is confined in this way it is possible to clip a third off from one of her wings so that she cannot fly. I don't practise this. There are a number of gadgets available to confine the queen for marking. I like the one which consists of a plastic cylinder, open at one end and with a mesh at the other. The queen is placed in the cylinder and confined there with a sponge piston, which can be moved up until the queen is gently trapped between the piston and the mesh where she can be easily marked. The advantages of this gadget, as I see it, are that you have the opportunity of retaining her in the cylinder, first while you find the marker pen and afterwards to allow the ink or paint to dry and you can move away from the hive into good light where you can mark her with confidence that she cannot escape.

Finding the Queen in Difficult Cases

There are times when you really do need to find the queen but everything is contriving against you. The colony is massive and the bees are decidedly tetchy. There are a couple of techniques that can help. The first one is to temporarily split the colony. Take the brood box and floor across to an unused hive stand at the other side of the apiary. Put the supers back on to a spare floor in its original position and replace the crown board. The foraging bees and hopefully the guard bees will return to the original site, and then after leaving the brood box for a few minutes you can recommence battle, now with the odds much more in your favour. A further technique is to split up the frames in the brood box. First remove three outer frames, ensuring that the queen is definitely not on them, and then reposition the eight remaining frames into four pairs with each pair separated from the next pair by a gap. Cover with the manipulation cloth or a crown board and leave them to settle for ten minutes. When you return the queen will almost certainly be between one of the four pairs of frames. It should be possible to detect which one as there will be more bees there.

If the queen has not been spotted then the presence of freshly laid eggs is a very good indication that you have a laying queen. It takes three days for an egg to hatch and begin its development as a larva. So, when you see eggs that indicates that you had a laying queen in the colony less than three days ago. In a balanced colony for every four frames of sealed brood there should be two of larvae and one of eggs. Seeing eggs is not always that easy and again a good pair of glasses can make a significant difference. The eggs are white and the size of the point of a needle and are placed at the bottom of the cell. It is sometimes necessary to stand with the sun behind your back and then adjust the position of the frame so that the light shines directly into the cells. It is much easier to see eggs in new comb, which the light shines through, than in old black comb. A torch can help. The egg when it is newly laid stands upright in the cell and gradually leans over as it approaches the time when it will hatch. Though individual eggs are more difficult to see than a queen, there are potentially lots of them and with a little experience you can anticipate where in the colony they are likely to be found. They are unlikely to be on full frames of honey, pollen or sealed brood or on frames of foundation that haven't been drawn out. They are likely to be within or on the edge of the brood nest and in fully drawn out and apparently vacant cells. In many ways the observation of eggs can be a quickest and more reliable way of determining the colony is queen right.

Unfortunately, the success or failure to find a queen and eggs in a colony does not absolutely determine whether or not the colony is queen right. During the period after a queen has hatched from the queen cell and before she starts to lay, there will be no eggs and yet there will be a young queen. Even after mating she will remain much smaller than a laying queen and much more likely to move quickly over the comb, and being a new queen will not be marked. She will also be still able to fly and can escape detection by taking to the air. And yet the colony is in a stable and viable state. The beekeeper needs to be patient and many inexperienced beekeepers panic at this stage. In most cases a new queen will not start laying until all the sealed brood produced by the old queen has emerged. Opening up the colony prior to four weeks after queen cells have been first seen can be counterproductive, but if the colony is examined and is calm, if there are polished used brood cells at the centre of the old brood nest and if the foragers are still bringing in pollen, then the indications are that all is well. You will notice that in this section I've been challenged into using a variety of qualifiers such as 'usually', 'more likely', 'in most cases'. There are no absolutes in describing the behaviour of honeybees, which is both a frustration and fascination of being a beekeeper.

We have just seen that the failure to see a queen and eggs does not mean that the colony is not queen right. It is also unfortunately the case that seeing a queen or eggs is not an absolute guarantee that all is well either. Don't be concerned that this is now far too complicated and you will never cope. There are many beekeepers that find these ideas difficult and I guess there are very few paragons of this craft that always read their bees correctly, and I am certainly not amongst them.

Drone Laying Workers

If a colony is left without a queen for more than four weeks, a small number of workers will start to lay eggs, the queen pheromone and pheromones produced by brood that inhibited this tendency now being absent. As these workers have never mated and are unable to mate, all the eggs will develop into drones and for this reason the condition is known as 'drone laying workers'. Once the condition is suspected it is easy to confirm.

1. There will be no queen
2. The eggs are laid on the side of the cells
3. Often there is more than one egg in a cell
4. The eggs are laid in worker cells but develop into drone brood
5. The eggs are not laid in a compact group, but at random across the comb

Drone laying workers are bad news. By the time that the condition is detected there are no new workers being produced and the population of workers is rapidly decreasing. The bees are generally dispirited and have ceased to forage with any energy. At an advanced stage the population will be dominated by drones that may hope to mate in a last desperate chance to ensure the survival of the genes of the colony. It is not possible to introduce a new queen, as the laying workers are themselves producing queen substance pheromones and the colony will kill any strange queen that is introduced. If the condition is detected early enough the non-laying worker bees may be saved and then unified with a viable colony, but as they are aging and probably few in number there is little benefit in doing it, except that you can then feel that you have at least done your best for them. To remove the laying workers, the colony should be carried to the far side of the field, at least a hundred metres from the apiary, and all the bees shaken out on the grass. A brood box with fresh comb is placed on the original site and will be repopulated with the flying bees which return to the original site. The laying workers, being heavier, are unable to fly back and perish on the grass. This small colony can then be unified with another colony with a viable queen. But it could be wiser to cull the colony entirely.

Drone Laying Queens

Once a queen has exhausted her supply of sperm in her spermatheca, all the eggs she lays will be drones. The colony is no longer queen right. This may occur at the end of a long and productive life, but in this case the workers will usually detect that this is about to happen and she will be superseded, decently dispatched and the life of the colony will continue. Drone laying queens are sometimes seen in early spring when the old queen starts to increase her egg laying rate. More often a colony has a drone laying queen when the young queen, on which its future hangs, fails to mate properly. The condition can be easily confused with having drone laying workers, but there are differences.

1. The brood is produced in large areas of compact drone comb
2. Single eggs are correctly laid at the base of the cell
3. There is a queen

But in this case the outlook is less gloomy, though again the population of workers will be depleting. The drone laying queen can be removed and a new queen introduced, giving the colony a new lease of life.

Assessing the Quality of the Queen

There are two main purposes of record keeping. One is to remind you what you did on your last hive inspection and the second is to enable you to form a view on the characteristics, good and bad, of the queen, so that a judgement can be arrived at as to whether the queen should be used to breed new queens. Desirable characteristics which you should consider are

1. The ability to survive in the environment where you keep your bees
2. Immunity from disease
3. Adaptations to deal with pests
4. Lack of aggression
5. Lack of the tendency to follow
6. Easy to manipulate by being calm on the comb
7. Fecundity of the queen
8. Able to produce a honey surplus
9. Reduced tendency to collect propolis

Unfortunately, it is not possible to have everything. Some adaptations to deal with pests may result in increased aggressiveness. To me it seems highly likely that immunity from disease could go hand in hand with propolis collection and I have seen some scientific confirmation of this. Fecundity of the queen does not necessarily correlate with the production of large surpluses of honey. Some bees can consume lots of honey as well as collect it. The choice of the most desirable characteristics will depend on your own aims as to what you want to achieve from your beekeeping and where you keep your bees. Aggressiveness in honeybees is not an absolute. Rather, there is a continuous spectrum, a range of temper from being pussy cats to being totally psycho. Overly aggressive bees are totally undesirable under any circumstance. Bees that require a little smoke and gentle handling may be quite satisfactory in an outer apiary, but in a small garden with close neighbours they could be a liability.

I tend to think that the ability of the honeybee colony to survive in the environment where they are going to live is the number one priority. This leads me to think that, in general, it is desirable to breed your own bees from your own queens that have the qualities that you most value. I do not favour buying in queens from abroad. I have bought some in the past and besides giving me a guilty conscience, I never was satisfied with their performance and felt that I had wasted my money. By contrast I have occasionally bought queens from UK breeders and some have been very satisfactory. But the best queens that give me the most satisfaction are those that have come from my own stock.

When raising new queens, it must be acknowledged that half of the genetics will come from drones that most beekeepers are simply unable to control.

I have no intention of talking further about raising queens besides what I've written about swarm control. It is too big a topic.

Introduction of Queens

There are several reasons why you may wish to replace the queen.

1. As a part of a swarm control policy it is common practice to replace queens after the second season. It is known that as queens age they produce less queen substance, the pheromone that inhibits the production of the queen cells, and so by having young queens the probability of swarming is reduced. Of course, if you have Carnican bees they will have swarmed in the first two months, never mind the first two years!

2. The queen is not performing. Occasionally you get a queen that seems quite content to maintain just two or three frames of brood

and will carry on indefinitely like this. Unless you want to use her in an observation hive, she needs to be replaced.

3. The genetic makeup of some queens can result in an increased susceptibility to some diseases like sac brood and chalk brood. Certainly, the best response to these conditions is to replace the queen. The immune system of insects is innate, not adaptive.

4. The temper of the colony is unacceptable. It is not necessary to kill all the bees in the colony but it is necessary to replace the queen. The change of temper affected by having a new queen can be remarkably rapid. Even before the offspring of the new queen have come to dominate the population, the new queen's pheromones will start to have an effect on the colony's temper.

5. As part of a breeding program, you may be wishing to run your bees with a single race of bees.

Each colony has its own colony odour, which gives each colony a unique identity. This odour is a complex combination of the pheromones, particularly from the queen, and odours from the unique mix of nectars that have been brought into the hive. The hive odour is continually being circulated throughout the colony as the bees groom each other and exchange nectar. The members of the colony use this odour to identify fellow members of the colony. This is an essential security measure to protect the valuable assets within the hive, an ID card to allow entry into the hive and protect the colony from bees from other colonies who may wish to rob the honey stores. Unfortunately, the ID card check also will apply to a new queen introduced into the colony, and if she hasn't the colony odour then there is a likelihood that she will be killed, regardless of whether the colony is queenless or not.

It is not possible to introduce a queen into a colony where there is already a queen. So, the first thing is to remove the queen that is being replaced. The colony needs to be left without a queen for several hours so that the new queenless state becomes known throughout the colony. However, this period should be less than twelve hours or otherwise the colony will begin to build emergency queen cells. If the new queen has been purchased, she will arrive in a plastic cage with a retinue of half a dozen workers. The retinue workers need to be removed. This can be done by releasing the bees from the cage inside a sealed clear plastic bag and then replacing the queen by herself in the cage. To introduce a queen the general principle is to place the queen in a cage and then place the cage between two frames at the centre of the colony. The cage allows the workers within the colony to touch and feed the queen through a grill or mesh, but they cannot harm her. In the course of a few days the pheromones of the new queen come to predominate in the colony at which point the queen can be safely allowed to leave the protection of the cage. The release of the queen can be brought about automatically by initially blocking the exit to the cage with fondant which is gradually eaten away until the queen can escape. After four or five days the cage should be checked to ensure that she has been released. The Butler cage remains the most popular introduction cage though there are now several plastic alternatives.

Even with the precautions described above, introduction of queens can be problematic, especially if the queen being introduced is of a different race. Queen introduction is usually more successful into a nucleus, which, once the new queen is established and laying successfully, can then be united with a larger colony from which the old queen has been removed.

Chapter 11

Feeding

Sugars

Honey is in the most part a super saturated solution of sugars. Sugars, along with fats and starches, are carbohydrates, the primary fuel for all living things including honeybees. Each gram of sugar has a calorific value of 4 kcal. There are a large number of different types of sugars, but most are multiples of the basic single unit sugars, monosaccharides, of which there are two, glucose and fructose. Glucose and fructose both have the same chemical formula, $C_6H_{12}O_6$, but differ in the way the molecule is arranged in space. The two monosaccharides have slightly different physical properties. For example, they alter the plane of polarised light in different directions and this accounts for their alternative names, glucose (dextrose) and fructose (laevotose). Rather surprisingly, they have different levels of sweetness, fructose being sweeter than glucose. This accounts for different honeys having different levels of sweetness. And their solubility in water is markedly different. At 20°C 83g of glucose will dissolve in 100gm of water. This compares to 375g of fructose.

The main sugar type found in nectar is sucrose, a disaccharide. Sucrose is the granulated white sugar which we buy from the supermarket. Commercially it is derived from either sugar beet or cane sugar. White sugar is, to all practical purposes, a pure substance, so there is no difference between sugar from beet and sugar from cane. With the addition of water and in the presence of the enzyme invertase it can be split into the two monosaccharides glucose and fructose. Sugar purchased in this form is known as inverted sugar. This is not usually available in a supermarket but can be bought from suppliers to the catering industry and suppliers to beekeepers. Other white sugars such as icing sugar or caster sugar are different physical forms of sucrose.

Sugar syrup, a solution of sucrose in water, is the usual feed given to honeybees. When you examine tables of solubility you will see that at 20°C, 200g of sucrose will dissolve in 100g of water. This produces a solution of 66%. In practice it is difficult to achieve this level of concentration, even using hot water. But even if you achieve a high level of solubility by using hot water, as the solution cools, you will find that some of the sugar crystallises out in the feeder and this can block the gauze or the narrow gap in the feeder so that the bees are unable to access the sugar solution. In practice 60% is a sufficient concentration. This is achieved by adding about 670ml of water to 1kg of sugar.

There is a school of thought that 1:1 concentration (one part sugar to one part water by weight) is appropriate if the feed is to be used by the bees directly, either to meet their immediate needs or to draw out wax and 2:1 concentration (two parts sugar to one part water by weight) if the intention is to produce honey stores. I'm not sure this complication is necessary and except for feeding weak colonies I use 2:1 solution.

Inverted sugar solution is increasing in availability and popularity. It is significantly more expensive than buying pure sucrose, but it has a number of advantages. Being already in liquid form it is immediately ready for use and it can be kept in containers for long periods without spoiling.

Methods of Feeding

In general terms there are three methods that beekeepers employ to feed their bees with sugar solution. The first is to allow the bees to draw the solution through a fine gauze which is set into the lid of an inverted sealed container full of syrup. The sugar solution is held in place by a partial vacuum above the liquid in the container. This is the principle which contact feeders use. These can be purchased from equipment suppliers and are available in different sizes.

They have three advantages.

1.   They are cheap.

2.   They can be placed directly over the brood nest so the bees can

access the feed without leaving the warmth of the brood nest. This makes contact feeders particularly useful for feeding in the early spring or for feeding small or weak colonies.

3. They can be filled at home in the kitchen or workshop, and then, covering the gauze with duct tape, they can be taken to the apiary. The feeders can then be tipped upside down and holding the container over the feed hole in the crown board, the tape can be removed and the feeder lowered into place.

They also have a number of disadvantages.

1. When placing the feeders on the hives it is easy to spill syrup, which can then lead to robbing.

2. The bees tend to propolise the gauze which reduces the effectiveness of the feeder.

3. Contact feeders are not easily replenished in the apiary without spilling some of the syrup.

Contact Feeder

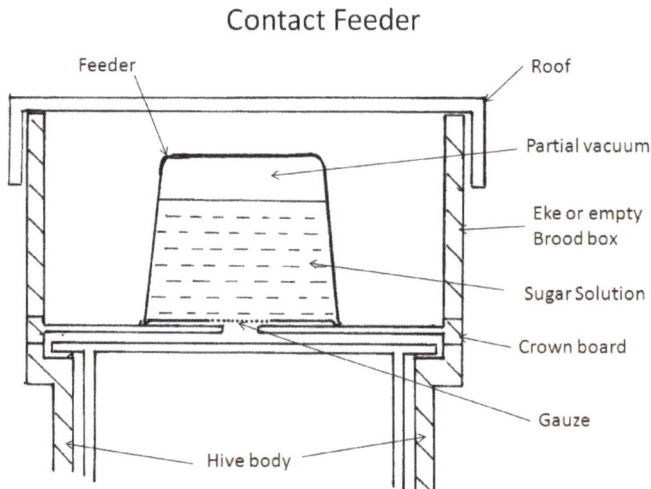

I own a few contact feeders but only use them for small or weak colonies.

The second method of feeding is by exposing a narrow surface of the sugar syrup to the bees within a space accessible from the inside of the hive. The bees can access the syrup by clinging to a vertical or sloping surface adjacent to the liquid surface. The feeding compartment is connected at the bottom to the main reservoir, which has a lid and so may be topped up in situ.

There are a number of ways in which this principle is implemented. The most popular is the rapid feeder. These are manufactured from plastic and are circular in shape. The bees access the feeder through a cone at the centre of the feeder. The cone is covered by an inverted beaker. The rapid feeders are placed over the clearer or feeder hole in the crown board, and then can be filled in situ. They are easily cleaned and can be topped up without spillage. Their capacity is 2 litres. A reasonably strong colony will empty the feeder in one or two days.

Rapid Feeder

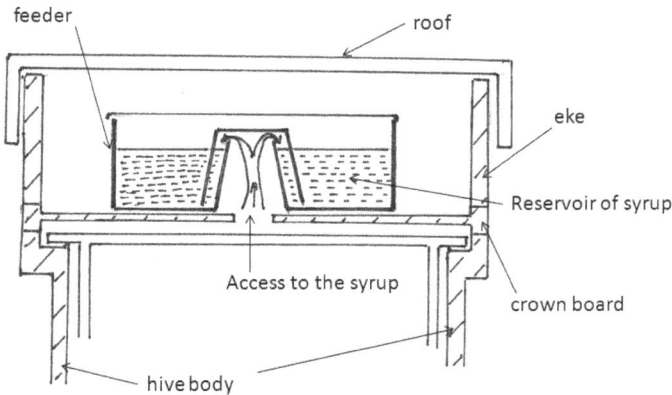

Miller Feeder

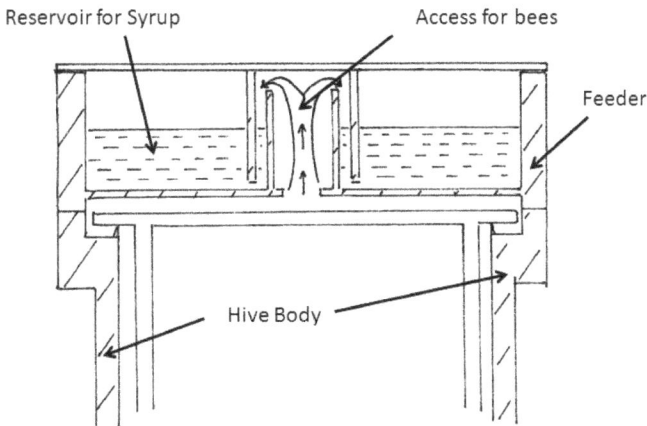

The same principle is used in the Miller feeder and the Ashforth feeder. These are usually made from wood. These feeders are both constructed with the same horizontal dimensions as the hive and are about 75mm deep, though this can vary. The feeding area is a slot stretching across the width of the feeder. The Miller feeder has the feeding slot positioned centrally so the bees can feed on both sides, while the Ashforth feeder has the feeding slot positioned at one edge. They both can contain about 12 litres of syrup. They are the best solution for bulk feeding in the autumn. Though equivalent in many ways, each has its own advantage. The Miller feeder has twice the length of feeding slot and so can feed a colony very quickly indeed. The Ashforth feeder can be positioned on the hive with the feeder slot at the lowest edge and so the bees can have access to all the syrup provided, wasting very little. Equipment stockists have available an economically priced plastic moulded version of the Miller feeder which fits within an eke. Its capacity is less than the standard Miller feeder. It can be converted to allow the bees to clean up the honey from cappings produced during the extraction of honey.

The third method is to place a float on a syrup reservoir, designed so that there is a narrow clearance between the float and walls of the reservoir. The bees are then given access to the float, and from the float can imbibe the syrup. This method is used in frame feeders, which are inserted into the brood box replacing a frame. They are often used in a nucleus box, and are particularly useful for feeding weak stocks.

If honey is to be fed to bees it should only be honey that has originated from your own bees. Imported honey may contain spores of EFB or AFB. Honey may be diluted to form a sugar solution and fed in one of the ways which has been described previously. Alternatively, it can be fed directly from a honey jar by replacing the jar lid with a lid with a gauze insert. The jar is then inverted over the feed hole.

If feeding during the summer and autumn, when possible, the feeders should be put on to the hives and topped up after sunset, reducing the frenzy that can result when a new supply of forage is located. As far as possible all colonies in an apiary should be fed at the same time.

Honey Consumption of a Colony

The beekeeper needs to have an understanding of the energy requirements of the colony. There are two main needs for energy, the basic metabolism of each individual bee in the colony and the energy required to raise brood.

The rate of energy consumption for metabolism for the individual honeybee varies with temperature. It is a minimum at 10°C, and a maximum at 19°C. Below 10°C the rate increases gradually as the temperature becomes lower. There is a rapid increase in the metabolic rate as the temperature increases from 10°C to 19°C and then it decreases gradually as the temperature continues to further increase. In particular it can be seen that the energy requirement for basic metabolism of the colony increases significantly as the temperature rises during the spring months, and at the same time the colony is producing brood at an increasing rate. To raise each new bee requires an expenditure of 142mg of honey.

The table below models the honey requirements of a typical colony as it goes through the year. The temperature profile given is the mean throughout the day for Yorkshire. Be aware that the figures shown are a mathematical model and do not relate to any given colony. Nevertheless, the table illustrates a number of important issues and analysis of the table illustrates the following points.

Analysis of Honey Consumption by a Colony of Honeybees

Month	Mean Temp degC	Colony Adult population '000s	Honey for metabolism mg/sec/bee x 0.0001	Honey for Metabolism / month kg	No of brood produced per month 000's	Honey for Brood / Month kg	Total Honey Required per Month kg
Jan	4	12	0.65	2.02	0	0.00	2.02
Feb	4	11	0.65	1.85	2	0.28	2.13
Mar	6	10	0.61	1.58	5	0.71	2.29
Apr	8	13	0.59	1.99	10	1.42	3.41
May	12	30	0.76	5.91	30	4.26	10.17
June	14	45	1.04	12.13	45	6.39	18.52
July	17	60	1.52	23.64	30	4.26	27.90
Aug	16	50	1.33	17.23	15	2.13	19.36
Sept	14	30	1.04	8.08	10	1.42	9.50
Oct	11	20	0.66	3.42	4	0.57	3.99
Nov	8	15	0.59	2.29	0	0.00	2.29
Dec	5	13	0.62	2.08	0	0.00	2.08

1. The table suggests that the total honey consumed by the colony for the year is over 100kg. This means a colony needs to produce 100kg of honey before it produces any surplus for the beekeeper.

2. According to the table, the honey consumed in the seven months over the winter, October to April, is 18.2kg, an amount that relates closely to the accepted amount (18kg) of stores that a colony of honeybees requires so that it is able to survive the winter. During the latter part of April, the colony is growing at a very high rate but the amount of forage available is increasing too, and so sometime during April the colony will change from being dependent upon its stores to being able to start to replenish those honey stores.

3. The table illustrates that a third of the recommended winter honey stores is likely to be consumed in March and April, at a time when the stores are becoming depleted and the colony should be increasing its rate of producing brood. Experienced beekeepers are aware that most colonies that succumb to starvation during the winter, actually die during these two months when it is easy to believe that the worst of the winter is past. In particular, it is strong colonies that are susceptible to starvation at this time. During this period colonies are using approximately 0.7 kg of honey each week, so if feeding is required the amount required is 0.7 litres of 2:1 sugar solution each week. If it is necessary to spring feed with sugar syrup then the feeding will need to continue until well into April, when there should be sufficient forage available.

4. During the months in the middle of the summer the colony consumes an enormous amount of honey. The model suggests that 28kg is consumed during July, the equivalent of two supers full of honey. In a single week the colony requires 7kg of honey. Of course, in normal circumstances we expect that during the summer the colony will be foraging energetically and bringing into the hive at least the amount of nectar required to produce the honey to meet its daily needs. But during periods of unsettled weather or when there is no available forage (the June gap) the colony may be unable to meet its day to day requirements. In the first instance the colony will simply use the honey that it has stored previously, but should the dearth continue over an extended period then this store can become exhausted and feeding may be required. The colony can

adapt to the situation to a significant extent by cutting back on brood production, but this is at the cost of a reduction in the strength of the colony later in the season, and will prejudice the ability of the colony to survive the following winter. To provide the full amount of the energy requirements a colony would typically need about 7 litres of sugar solution each week. However even in a dearth at that time of year the bees will be managing to bring in to the colony a significant proportion of their requirement, so this extreme amount would not normally be required. But the beekeeper needs to be continually aware of the amount of forage that is available, the weight of honey stores in the colony, and be ready to feed when necessary. Once a colony comes under nutritional stress, it will be vulnerable to other diseases.

Autumn Feeding

A colony of bees requires 18 – 20 kg of honey to survive the winter. Many beekeepers maintain that honey produced from nectar collected during the summer is by far the best for the purpose, much better than the honey produced by feeding sugar syrup. There is some merit to this argument. Natural honey contains small quantities of pollen, which itself contains protein, vitamins and trace elements, none of which are present in sugar syrup. On the other hand, there can be downsides to relying solely on natural honey for winter feed. Some types of honey, particularly heather honey, contain higher proportions of water and this can lead to outbreaks of dysentery during the long winter confinement in the hive, and this in turn can promote the spread of nosema. Other honeys, such as those based upon nectar from cruciferae, for example oil seed rape, readily granulate and the bees can have difficulty utilising it without access to water. The bees may use condensate from within the hive, but in a well ventilated hive this will be insufficient.

The necessity to treat for varroa requires the introduction of chemicals into the hives. Most of the substances now used can be found in some natural honey, but quite rightly beekeepers should try to avoid treating for varroa when there is a risk of contaminating in any way the honey that will be harvested for human consumption. At the same time autumn feeding should be completed by the end of September. As a result of these constraints most summer honey needs to be harvested in the second or third week of August giving four clear weeks during which the varroa treatment can be carried out, followed by a period of two weeks for feeding.

After the honey is removed, small colonies can be reduced to a single brood box, large ones to a brood box plus a super. In the latter case the queen excluder should be removed. During September the area of brood is now rapidly contracting and the comb no longer required to raise brood will be filled with honey stores. After harvesting the honey, the colony has eight weeks until mid October to continue to forage to produce honey for the winter stores. In the last two weeks before the end of September the honey stores in all colonies will need to be assessed. In some years, by the end of September, many colonies will have collected sufficient stores to survive the winter, but when there is less than 18kg, then the shortfall needs to be made up by feeding sugar syrup.

Assessing the weight of honey in the colony can be carried out by counting the number of frames of honey in the hive or by weighing the hive. A BS brood frame can contain about 2.5kg of honey, a BS shallow frame about 1.5kg of honey. From those figures it can be seen that there needs to be between 7 and 8 frames of honey in a brood box. An experienced beekeeper will be able to judge the weight of the hive by hefting, by lifting one edge of the hive off the hive stand. The less experienced or more scientific beekeeper may prefer to use a spring balance to lift one edge, and an estimate of the weight is obtained by doubling the reading. A colony in a single brood box containing 18kg of honey will have an overall weight of about 29.5kg. The shortfall in the winter honey store is the difference between the actual weight and this target weight of 29.5kg. The weight of the additional honey that the bees need differs from the weight of sugar and the sugar syrup required to produce it. A kilogram of granulated sugar, which is in a 60% solution has a volume of 1.2 litres, and this is equivalent to about 1.2kg of honey. The table below gives the amounts of syrup solution required for autumn feeding a single brood box colony with different weights.

Weight of Hive (kg)	No of litres of sugar solution required
14	15.5
16	13.5
18	11.5
20	9.5
22	7.5
24	5.5
26	3.5
28	1.5
30	0

It can be seen that in some circumstances a great deal of sugar is required. Buying a couple of bags of sugar from the supermarket can be insufficient. Beekeepers need to buy their sugar in 25kg bags.

Even though the bees may continue to forage natural sources of nectar, in particular ivy, well into October, the uncertainty of the weather demands that the feeding of colonies is carried out before the end of September. The bees need to be active within the hive to be able to process the sugar solution, and as the temperature falls during October, the bees will spend more of their time in the cluster. In addition, the bees need to be able to produce invertase, an enzyme that breaks the disaccharide sucrose into glucose and fructose. The invertase is produced by the hypopharyngeal glands in the head of the workers and requires pollen to synthesise it, and pollen becomes less available as the autumn progresses. Late feeding can be facilitated by using inverted sugar, which is already in the form of monosaccharides, but this is more expensive. Leaving sugar solution in feeders on the hives over long periods can also cause problems, as it may ferment, and this can lead to dysentery.

Winter Feeding

Despite one's best efforts to provide a colony with sufficient stores in the autumn, it is essential that the state of the stores in each colony is reassessed during January, February and March. Although in normal circumstances a properly provisioned colony will not starve during the winter, unusual conditions may have occurred that have accelerated the consumption of the stores collected during the autumn, such as exceptionally mild weather causing the bees to fly excessively, or exceptional cold. The rate of metabolism of a colony in a cluster is low and in December the stores are being used at just over 2kg per month. This rate of metabolism further decreases in January and February as the population of bees within the cluster decreases and later on as the temperature of the environment increases. On the surface it would seem that the 18kg of stores should be more than sufficient to allow the colony to survive the winter. However, in January the colony will slowly start to produce brood and this places an additional and increasing demand upon the stored honey resource. It is in March and April, as brood production accelerates, that the real risk of starvation occurs.

Should additional feed be required during the winter months, sugar syrup cannot be used as the bees will have difficulty processing it. Instead fondant is the better option. Fondant for bees can be bought prepacked in sealed plastic bags that just need to be slit open before being placed over the feed hole of the crown board. Alternatively, cheaper but less convenient, it can be bought in 12.5kg slabs that need to be cut into 1-2 kg blocks and these blocks placed in containers, such as margarine containers or the tin foil containers used to hold takeaway meals. A 2kg block should be sufficient to provision a hive for a month. The container can be inverted over the hole in the crown board, or even placed directly on top of the brood frames, surrounded with an eke to make the space below the crown board. There are occasions when I have used fondant when I suspect that it was not absolutely necessary, and I was being over cautious. It is cheaper and more effective to feed sufficient sugar solution in autumn, than rely on emergency winter feeding of fondant. But fondant must be offered if the bees are on the point of starving.

Feeding in March and April is different. At this time of year, a large colony with a fecund queen could be producing considerable amounts of brood and as a result using what the remains of winter stores of honey at an increasing fast rate. A small amount of forage, providing both nectar and pollen, is available once the snowdrops and crocuses are in flower, though these sources are intermittent and unreliable. Once the daytime ambient temperatures reach 10°C, the activity within the hive will reach a level so that the colony will be able to process liquid syrup feed.

Protecting the integrity of Honey Produced

The beekeeper is duty bound to ensure that honey that the bees have produced as a result of having been fed sugar is not eventually extracted and used as honey for sale for human consumption. Any excess frames of such honey should be removed from supers at the start of a honey flow, stored and then reintroduced into the hive in the autumn as a contribution to the winter stores.

Chapter 12

Keeping Healthy Bees

This is what it is all about. Whether you are keeping pigs or cattle or looking after your own children or grandchildren, if you keep them healthy then they thrive. And so, our first priority as beekeepers is to keep our bees healthy. This chapter is not about treating bees once they are sick - it's about endeavouring to ensure they don't get ill in the first place.

At one time I considered keeping chickens and in preparation borrowed a book out of the library to read up on the subject. Two thirds of the book was devoted to describing a whole series of horrible diseases that my chickens might catch, accompanied by grizzly pictures. I was discouraged. I don't want to do the same here, but to stress that with simple measures to isolate your bees from disease, changing comb regularly and ensuring that they are well nourished, then beekeeping need not be a continual battle with disease.

Isolating your Bees from Disease

Beekeepers are the main vectors responsible for the spread of disease amongst honey bees. This is illustrated by taking the example of varroa. Varroa will, even without human intervention, spread from colony to colony. To do this they need to have physical contact between members of an infected colony and a neighbouring uninfected colony. Varroa mites have no means of moving more than a few centimetres without the aid of their honeybee hosts. Robbing and swarming are the natural mechanisms for spreading varroa from colony to colony. It is difficult to envisage these or any other natural mechanisms spreading varroa by more than a couple of miles each year. But varroa spread from Devon, in 1992 to Nairn in northern Scotland in 2008, a distance of over 550 miles in 16 years. The reality is that it spread at an average speed of almost 40 miles a year. It is a sad fact that the rate of dispersal had been multiplied twenty fold by the action of beekeepers, through carelessness, through lack of thought or through sheer irresponsibility. As far as varroa in the UK is concerned it is too late, but there are other new pests and diseases that we may need to confront in the future.

In the natural state, colonies of honeybees are isolated organisms. The members of one colony have limited contact with the members of another colony. The whole concept of an apiary is artificial. Honeybees are not by nature herd animals. Indeed, when honeybees swarm they do their best to disperse as much as possible. Feral colonies in a wood have a population density usual less than one colony per square kilometre. As a result, in the wild, honeybees had no requirement to evolve an immune system that would react quickly to new diseases. They do have an immune system but it is designed to protect the bees within isolated colonies.

Keeping colonies in apiaries is unnatural. We need to recognise that and mitigate the risks associated with this practice. The first measure is to ensure that we do not keep too many colonies in one apiary so that the colonies are competing with each other for forage. Exactly what number is too many is open to debate and will depend on the amount of forage all year round that is available in the area. I have restricted my colony numbers in any one place to eight, though, whereas I'm sure that there must be a limit on numbers, this self-imposed limitation is somewhat arbitrary.

As beekeepers we need to use barrier nursing techniques on our colonies. We need to keep our bees so that, to a large extent, we preserve the isolation between one colony and another, and more importantly between one apiary and another. It is a sad fact that this simple principle is broken frequently and in many ways.

1. We collect swarms from a wide area and bring them into our apiary.
2. We combine colonies that we consider are too weak to prosper by themselves.
3. We move our bees about the countryside in search of abundant sources of nectar.
4. We use the same tools and wear the same clothes and gloves when

handling different colonies in different apiaries.

5. We buy and bring into our apiary second hand equipment.

6. We buy or inherit colonies of honeybees from doubtful sources.

7. We allow robbing by colonies within our apiaries.

I'm not suggesting that all these activities can be entirely removed from beekeeping. That is not realistic. Some are too deeply engrained, but we do need to be aware of the risks and then take measures to reduce those risks as much as we can.

Collecting Swarms

Though a swarm will generally be less infested with pests and disease than its parent colony, swarms will still carry with them pests and pathogens from the old colony. Varroa in the phoretic stage of their life will stay on the thorax of their host, EFB and AFB spores can be carried in the honey within the workers honey stomach, viruses are carried in the haemolymph and nosema in the gut. Any swarm of unknown origin should be treated with suspicion. The most frequent way of bringing EFB or AFB into an apiary is through a swarm. Therefore, it is a good idea to have a small out apiary to act as a quarantine area. Swarms should be put on to fresh foundation, and after a couple of days fed with sugar solution. The delay in feeding the swarm is to try to ensure that the honey brought in the honey gut is fully utilised in building wax, rather than stored. After four or five weeks, when the brood is properly established, a full brood disease check should be carried out. Only then, if all is well, should the swarm be moved to the main apiary.

Establishing a swarm on fresh foundation gives an ideal opportunity to largely banish the phoretic varroa by using the oxalic acid trickle treatment.

It is tempting, in the heat of the moment, to combine small swarms together, in order to reduce the amount of equipment required. A better policy is to establish them in a nucleus box to start with and then combine them, if necessary, after the brood disease check and the quality of the queen has been fully assessed. It is a real risk to combine a swarm from an external source with an existing small colony in the home apiary.

Migratory Beekeeping

Migratory beekeeping is without doubt one of the most significant causes of the spread of honeybee pests and diseases. Unfortunately, bee farmers rely upon the productivity of their bees to enable them to make a living and feel that they have no option but to practise migratory beekeeping. In Yorkshire bees are regularly moved to the heather moors in August and to the borage fields in June. I personally would not practise migratory beekeeping without an absolute assurance from the farmer or estate manager that my bees on the moors or on a borage crop were isolated from the bees of other beekeepers by at least a quarter of a mile. The cost of honeybee disease is borne by us all, whereas the risk and associated profit from migratory beekeeping is taken by a small number of beekeepers.

Second hand Equipment

Beekeepers, not just those from Yorkshire, like a bargain. It follows that there is a healthy market in second-hand beekeeping equipment. Every spring there are well attended auctions and if you have the misfortune to die while still actively beekeeping, then, hardly will the first clod of earth be thudding down on your coffin before your heart broken partner will be cheered by the suggestion of a lucrative sale of your bees and equipment. Any bees bought at auction should be certified as having been checked for disease by an experienced beekeeper, other than the vendor. If second hand equipment is bought it must be sterilised before being used. The easiest method is to scrape them clean of wax and propolis and then use a blow torch flame to scorch all surfaces, taking particular care to direct the flame into corners and crevices. Alternatively, the equipment can be fumigated.

There are some things that should not be bought second-hand such as frames, combs and gloves.

Apiary Hygiene

The regional and seasonal bee inspectors give us all an excellent example of the proper apiary hygienic practices that we should follow, and it is worthwhile asking for your bees to be inspected simply in order to see the inspectors at work and learn from the techniques that they use.

Leather gloves can carry pathogens from one colony to another. Within your own apiary this is not an issue, but it is bad practice to wear your leather gloves in another beekeeper's apiary. The solution is to use disposable nitrile gloves. There are other advantages to using nitrile gloves, the main one being that you are automatically encouraged to handle the bees more sensitively. In order to prevent the bees stinging my wrists I wear gauntlets that cover the forearm, wrist and upper part of the hand. The overalls or smock should be washed frequently.

In a similar way the hive tool and other pieces of equipment pose a similar small risk. After each hive inspection all remnants of wax should be removed and if you intend to visit another apiary, equipment can be sterilised by using a blow torch or placing in sterilising liquid, such as a solution of washing soda. It is good practice to have a container of washing soda solution available in the apiary.

Robbing

In the last few weeks of the summer the bees are desperate to ensure that they have sufficient stores of honey to enable them to survive the winter, and this desperation will lead them to rob neighbouring colonies if the circumstances allow. Robbing allows pests and pathogens to be transferred from colony to colony, not just from the robber to the victim, but in the opposite direction as well. The weaker colony could be weak for a reason, and the cause of that weakness can be transferred to the aggressor colony. Once robbing is underway it is difficult to control. The best policy is to ensure that it is not allowed to start.

Under normal circumstances a colony is able to defend itself. It is obviously not a sustainable strategy for bees to expend their energy robbing honey stores from each other. That would be the path to ruin for all colonies involved. However, it is a sustainable strategy for a strong colony to rob the stores from a weaker one, even if it results in the weaker colony perishing, as it surely will. One of the roles the workers perform is to act as guards at the hive entrance. Foragers from other colonies will occasionally test these defences, and make their aggressive intentions clear by zigzagging in front of the entrance. The guards of a strong colony will be alerted and should the robber try to enter the colony a fight to the death will occur. Normally the potential robber, having assessed that the colony is well defended, goes elsewhere, possibly deciding to earn an honest living.

Once robbing starts the only solution is to move the weaker colony to an out apiary, and then reduce the entrance as much as possible. However, robbing is not inevitable, and there are several steps that can be taken to stop it occurring in the first place.

1. Use a reduced entrance. Traditionally the entrance to the hive served two purposes. Besides allowing the bees to leave and return, it also was a duct for ventilation, and as a result, during the summer months, it extended across the full width of the hive. Now that many beekeepers are using mesh floors, ventilation is through the mesh base and so the entrance can be much reduced right throughout the year. I find that an entrance width of 100 – 150 mm, and height of 9mm is more than sufficient. The reduced width makes it easier for the guard bees to defend. The entrance to feral colonies is often much smaller than this. However even if a wide entrance is being used during the peak of the summer, it should be reduced in August, especially for smaller colonies.

2. Keep your hive boxes in good condition so that there is no access to the inside of the hive except through the hive entrance.

3. Apiary layout. Robbing can be initiated by foragers drifting almost by accident into neighbouring colonies. Therefore, apiaries should be laid out so that the position of each colony is readily recognised by the returning foragers. Straight lines of hives should be avoided. There should be a designed randomness to the way the hives are arranged. Landmarks, for example shrubs, can be introduced.

4. Apiary Hygiene. Robbing can also be induced by poor apiary hygiene, in particular leaving scraps of honey coated comb that

have been scraped away during manipulations lying on the ground or on hive roofs. I recommend having a plastic container with a lid at hand in the apiary into which these scraps can be placed.

5.   Wasps. From the middle of August to the end of September apiaries can be targeted by wasps. Once a colony becomes a prey to wasps it becomes vulnerable to robbing from honeybees as well. Sometimes it is possible to track the wasps back to their nest and destroy them there. Alternatively, wasp traps can be set up about the colony and in this way reduce the population of these invaders. A wasp trap can be made simply, half filling a jar with sweetened, flavoured water and drilling a 6mm hole in the lid. In a week the jar is full of dead wasps.

6.   Feeding. Feeding should be carried out at dusk, when the foraging of the day is completed. Care needs to taken not to spill the sugar solution. Under no circumstances should honey be fed to honeybees except if it is sourced from your own bees. Most honey sold in this country is imported from overseas, often from countries where there is little control over the health of honeybees.

Nutrition

Honeybees have two main sources of nutrition. Nectar provides energy and pollen provides protein, vitamins and trace elements. Energy is required at all times, and so to overcome the lack of nectar during the winter months the honeybee superorganism has learnt to store nectar in the form of honey, and these stores must be sufficient to enable the colony to live throughout the winter and through times of dearth in the summer. On the other hand, protein is only required when the bees are producing brood. Honeybees store pollen in small quantities close to the brood nest and winter bees have fat bodies, which contain protein, in their abdomens, but when the colony starts raising brood it is essential that there is new pollen available from the local flora. The building blocks of protein are amino acids. There are about twenty amino acids. Of these, ten are essential amino acids which the honeybees (and human beings) cannot synthesise. For the record these are isoleucine, leucine, lysine, methionine, phenylalanine, threonine, tryptophan, valine, arginine and histidine. Pollen is a remarkably good source of protein. But pollen from different plants do not all contain a similar mix of amino acids and research has shown that honeybees are healthier if they are using a mixture of pollens. Even when there is a vast preponderance of one type of pollen available, such as when the hives are placed next to a flowering field of oil seed rape, it can be seen from the pollen loads that the bees are accessing several other types of pollen in significant quantities.

The beekeeper has some limited influence on the pollen which is available from the local environment. The beekeeper chooses where he sets up his apiary. If the apiary is on the beekeeper's own land then the beekeeper can choose the plants that he grows, possibly preferring a wild flower meadow to a manicured lawn or a willow tree to a flowering cherry. The beekeeper, possibly through his association, can suggest that farmers use clover mixes for their conservation strips, can try to persuade local authorities to bear honeybees in mind when sowing road verges and designing parks and by speaking to local groups such as the Women's Institute make the general public aware of the importance of designing gardens with bees in mind.

It is also possible to supplement the pollen collected by the bees, by feeding substitutes or feeding back pollen collected at times of plenty. There are a number of commercially produced pollen substitutes available. They are expensive and as I have never used them, I can make no judgement on how effective they are. I have tried to formulate substitutes from recipes in the beekeeping literature but found difficulty finding the ingredients. However, I have fed, in early spring, small patties of pollen collected in a previous season and the bees were happy to use them. Though my apiary is not blessed with tremendous flows of honey, I do seem to have sources of pollen throughout the year, from hazel in early spring to ivy in late autumn. I look back with some self satisfaction at my decision to plant willows and hazel on waste land close to the apiary twenty years ago.

Although a strong colony will normally have sufficient honey stores to enable the colony to live through the natural fluctuations in the availability of nectar, there will be occasions when the beekeeper must be prepared to intervene by feeding. The energy consumption of a typical colony has been discussed in the previous chapter. When there is a dearth the bees (with the exception of Italian) will, to some extent, adjust their energy consumption by cutting back on brood production. But the beekeeper needs to monitor colony weights during the summer when the weather is poor. Once a colony comes under stress by having insufficient nutrition then it will become vulnerable to disease.

Changing Comb

Changing comb on a regular basis is becoming a standard part of beekeeping practice and it is something I recommend. By preference feral colonies of honeybees raise brood in new comb. It is sometimes said that a feral honeybee colony occupying a given site, such as a hollow tree, lasts indefinitely. This is not true. It has been shown that on average, a colony of feral honeybees occupying a given site lasts for less than two years. When a colony dies out, wax moths move in to the cavity and remove the old comb, along with the pathogens it contains. When beekeeping was first practised using straw skeps the brood would be raised in comb that had to be rebuilt each year. The invention of moveable frame beekeeping changed that. The queen was confined to a specific part of the hive, the brood box, and it became possible and normal practice to use the same brood comb over a period of many years. When I started beekeeping, I overheard established beekeepers boasting about the number of decades that a particular brood comb had been used. In those halcyon days maybe, the risks were less. It is no longer regarded as good practice to retain brood comb for more than two or three years, and indeed changing comb on an annual basis is becoming accepted practice.

There are a number of reasons for changing comb.

a. It removes pathogens from the colony. Pathogens that are commonly present in the wax of the brood comb include nosema spores, EFB and AFB spores, chalkbrood.

b. During the course of a season each comb can be used and reused up to eight times. Each time a cell is used it is cleaned by the workers and lined with propolis and faecal matter expelled from the larvae. As a result, the internal dimensions of the brood cells are marginally reduced at each cycle and this may affect the size of subsequent larvae.

3.	It is much easier to spot eggs in new comb.

Old comb gradually becomes damaged and less area is available on each frame for brood. I try to maintain my colonies on a single national brood box. This makes inspections much easier to manage than some of the alternative hive configurations. If you adopt the brood and a half method, this will entail looking through twenty-two frames rather than eleven. 14 by 12 or commercial frames are uncomfortably heavy, especially for ladies or older beekeepers. The reality is that, for all but the most prolific of queens, there is potentially ample space in a single National brood chamber. Most authorities suggest that at the peak of the season, the queen can lay, on average, about 1500 eggs each day. If this rate was consistently maintained for 40 days it would result in a colony approaching 60,000 bees, which most beekeepers would consider to be quite a strong colony. Worker brood takes 21 days to develop from laid egg to emerging as an adult, so, allowing an extra three days for the empty cell to be prepared for receiving another egg, the total brood cell requirement is 1500 x 24 = 36,000. On each face of a BS deep comb there are about 2300 cells and so the total number of cells on the eleven frames in a National brood box is 2300 x 22 = 50,600. It can be seen that the number of brood cells required is about 70% of those available. Of course, not all cells are available for laying and as time goes by this availability does decrease. The bees cut holes through the comb, areas are used to store pollen and honey and parts of the comb are damaged beyond repair. And so, this is the fourth reason for regularly changing comb – that is to ensure that a large percentage of the brood comb area is available for the queen to lay in. Because, for many years, I have renewed comb on a regular basis, I expect and see brood laid almost across the comb, with just a small margin. In these circumstances I rarely get brood occupying more than 18 or 20 sides of comb.

It is my experience that once a colony has undergone a comb change, it is less likely to swarm in the weeks immediately following.

'A one year old brood comb, much reduced in area and partially pollen bound'

There are a number of approaches to changing comb. Many beekeepers have a routine of replacing three or four frames each year. To do this, three or four combs are selected for replacement during the first spring inspection and marked with an indelible pen or coloured drawing pin. These combs are gradually moved, during the course of the next few weekly inspections to the edge of the brood box and when they are clear of brood they are replaced with a frame of new foundation. This process is not altogether straight forward. Altering the relative positions of the brood combs within the brood nest disrupts the build up of the brood, and by spreading the process over several weeks it does require discipline to ensure that it is carried out correctly.

I now prefer to completely renew comb after it has been used for a full year, during which it will have undergone at least eight cycles of brood. In order to monitor the age of comb, when a frame is made up with fresh foundation, the top bar can be marked with a coloured cross using indelible ink, using the same colour code as queen marking.

There are several procedures to achieve a full frame exchange and here I will describe three of them in detail. Though they all have the same basic purpose, they are not entirely interchangeable.

The Shook Swarm procedure

The Bailey Frame exchange variant 1

The Bailey Frame exchange variant 2

The Shook Swarm Method

The shook swarm method could be said to mimic the process a colony undergoes when it swarms and re-establishes itself on new comb in a different site. But it differs from a natural swarm in a number of respects in that that there is no division of the colony, the colony remains on the same site and the colony is provided with foundation by the beekeeper. The NBU bee inspectors now prefer this method over using antibiotics as a treatment for a colony with EFB.

The basic steps in the process are

1. The colony is moved a metre or so from its original site.
2. A new floor and brood box containing frames with fresh foundation are placed on the original site.
3. The frames from the original colony are taken one at a time and the bees shaken or brushed into the new hive.
4. When all the bees are transferred, a crown board and roof can be placed on the new brood chamber
5. The colony should be fed with between two and four litres of 2:1 sugar syrup.
6. The old comb is removed from the apiary and then burnt or rendered down.

The above will, as often as not, work, but things can go wrong. There is a significant risk that the colony may abscond, as, for a while, it no longer has any assets to protect or maintain. There are two accepted approaches to reducing this risk, by preventing the queen leaving with an absconding colony. A queen excluder can be placed between the floor and the brood box or the queen can be placed in a cage confined with a plug of fondant. The queen excluder should be removed or the queen released after two or three days once the colony has started to draw out the comb. Caging the queen requires that the queen can be found before the manipulation begins. The presence of the queen excluder makes it more difficult for the bees to bring pollen into the colony.

Shaking the bees from the frames requires care. One approach is to remove two or three frames from the centre of the new box and shake the bees into the space created. This can result in a cluster of bees on the floor of the hive which may be crushed and damaged when the frames, that were removed, are replaced. The alternative is to create a barrier around the top of the hive with an empty super or a couple of ekes and then, holding the frames diagonally across the frames, shake the bees on to the top of the frames. The bees will quite quickly migrate downwards between the frames and a little smoke can hasten this process.

The shook swarm method has the obvious disadvantage that the colony loses all its brood and stores, which it might be feared would set back the development of the colony. The colony can be fed sugar syrup to provide an equivalent source of energy, and it is usually observed that colonies that have undergone a shook swarm will re-establish themselves very quickly, in the same way that a swarm will do.

This disadvantage is outweighed by the benefits. The loss of brood means that all pathogens associated with the brood are removed, including the varroa that are in the reproductive stage in the sealed cells. The method also presents the beekeeper with the opportunity to treat the colony with oxalic acid by the trickle method in the first few days after the shook swarm to destroy varroa which are in the phoretic stage. As a result, the shook swarm method is one of the most effective weapons in the battle against varroa.

The shook swarm method is not generally suitable for small colonies. Small colonies have difficulty raising an area of the foundation to the brood nest temperature so that the bees can draw out comb and the queen can resume laying. Nor is it suitable for colonies suffering from nosema. By the very nature of the procedure there can be collateral damage to some bees. When nosema is present it is absolutely essential that bees are not crushed, killed or stressed in anyway as the hygienic behaviour of honeybees in cleaning away dead bees will result in the spread of the disease.

Bailey Frame Change

The Bailey Frame Change is often thought of as simply a less traumatic alternative to the shook swarm method. Though the end result of moving the colony on to fresh comb is the same, there are significant differences as to when and where it is applicable. In particular, the Bailey Frame Change cannot be regarded as part of varroa control, but, on the other hand, it can be used to treat a colony infected with nosema.

This manipulation needs

a. An eke adapted so that in the centre of one side there is a gap of about 100mm wide and 10mm high that can act as an entrance to the hive

b. A sterilised brood box with a full complement of frames containing foundation and for variant 1, including a small number of frames of drawn comb.

c. Four dummy frames or wide dummies

d. A queen excluder

There are two variants to the Bailey Frame Change procedure.

Bailey Frame Change Variant 1

The first of these is used to transfer bees from comb contaminated with nosema spores. Colonies infected with Nosema necessarily will be weak and are unlikely to have a super. They must be treated with great care. If they are stressed the infected bees may defecate, and the house bees will attempt to remove the faeces infected with nosema, and spread the disease further. A similar thing will happen if bees are crushed or damaged in any way. This variant requires at least two or three frames of sterile drawn comb

On day 1

a. Find the frame with the queen and place it in the new brood box between at least two frames of drawn comb.

b. From the original brood box, remove all frames that are clear of bees, close up the remaining frames into the centre of the original brood box and confine what remains between two dummy frames.

c. On the original brood box place a queen excluder and then the Bailey eke with the entrance directly above the original entrance. And then on top of the eke place the new brood box, which now contains the queen. Add frames of foundation so that the number of frames in the upper brood box is the same as the number of frames now in the original hive, once again confining them between two dummy boards, mirroring the configuration in the lower brood box.

d. The entrance in the hive floor should be closed with sponge. The bees will quickly learn to use the new entrance in the Bailey eke. Place a new crown board on the new brood box and then feed, about 5 litres of 1:1 syrup would be about right. It is recommended that a contact feeder is used rather than a rapid feeder, as this gives the bees easier access to the feed.

On Day 5 or 6

By this time, the queen should have started to lay in the drawn comb she was given, and the workers will have started to draw out the comb on the frames of foundation. The old frame on which the queen had been found should be removed, gently brushing off the bees, and if necessary, the queen, into the upper brood box, and then that frame with the old comb can be replaced in the lower brood box. The frames in the upper brood box should be closed up and additional frames of foundation can be added to the upper box if necessary.

On day 21

In theory all the worker brood in the lower brood box will have emerged. In fact, it is often the case that there is still some sealed worker brood left, as the period taken for the pupa to develop varies slightly and can be extended by the brood nest temperature being below the optimum . Now is the time to remove the lower and original brood box, the queen excluder and the Bailey eke and place the new brood box on a floor, which should be replaced with a new sterile one. The entrance in the floor should now be open. If necessary, be prepared to add frames of foundation to the new brood chamber at his time and over the next few weeks as the colony expands through the spring and summer. Any bees on the frames from the original brood box need to be transferred into the new brood box. This can be achieved by placing the original brood box still with the bees over a clearer board above the new brood box for 24 hours. Finally, the old frames, as they are probably infected with Nosema spores, should be burnt.

Bailey Frame Change Variant 1

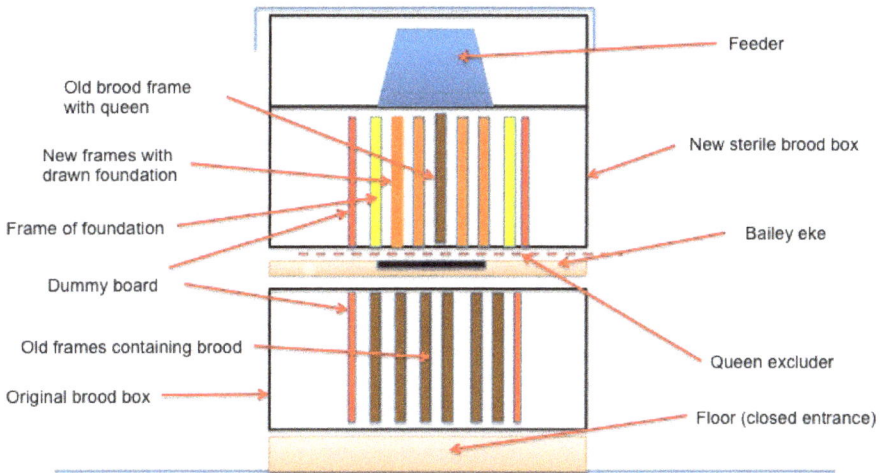

Bailey Frame Change Variant 2

This is applicable where the colony is thriving and healthy and the only purpose is to renew the comb. Unlike the shook swarm method there is no loss of brood, but it cannot be regarded as a treatment for varroa. This is a procedure I would aim to start the second or third week of April as the oil seed rape is coming into flower. This date can vary across the country and from year to year. It is better to carry out the procedure before the colony begins to produce drone brood in significant quantities as drones end up being trapped in the lower brood box.

About a week before the procedure is to be carried out, place a brood box, complete with a full set of frames of foundation, above the brood chamber. Take care that the foundation is placed accurately in the frames, without warps or curves, so that the bees are not encouraged to draw any wild comb between the sheets of foundation. Do not separate the two brood boxes with a queen excluder. Place a feeder over the hive with about four or five litres of 2:1 sugar syrup which should encourage the bees to draw out comb on the foundation.

Once the feed is exhausted the Bailey frame change can be set up. The queen needs to be placed on a newly drawn comb in the upper box. If you gave her a good talking to, maybe she will be there already and have started to lay. Then

- Place a queen excluder on the bottom and original brood chamber
- On top of the queen excluder is placed a Bailey eke, with the entrance facing in the same direction as the entrance in the floor
- Above the eke is placed the new brood chamber, containing drawn comb and the queen.
- The entrance of the floor is closed.

Setting up a Bailey Frame Change

A colony undergoing a Bailey Frame Change

Bailey Frame Exchange

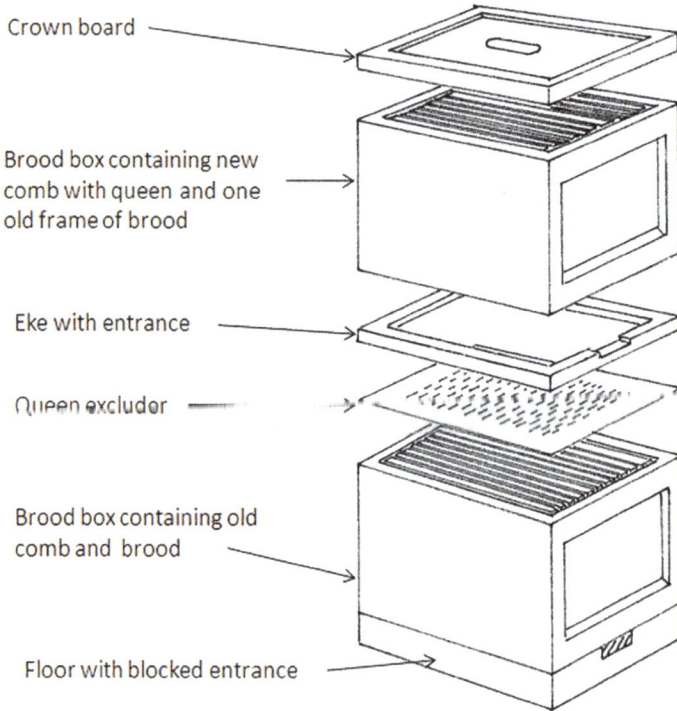

Crown board

Brood box containing new comb with queen and one old frame of brood

Eke with entrance

Queen excluder

Brood box containing old comb and brood

Floor with blocked entrance

Over the following three weeks the bees will use the upper entrance in the eke, the queen will establish an area of brood in the new comb in the upper box and brood in the lower box will develop and emerge as adult bees. After this period the lower box can be removed along with Bailey eke and the queen excluder. The entrance in the floor is then opened. Any bees in the lower brood chamber are shaken into the upper box or just outside the entrance.

Bailey Frame Change Variant 2

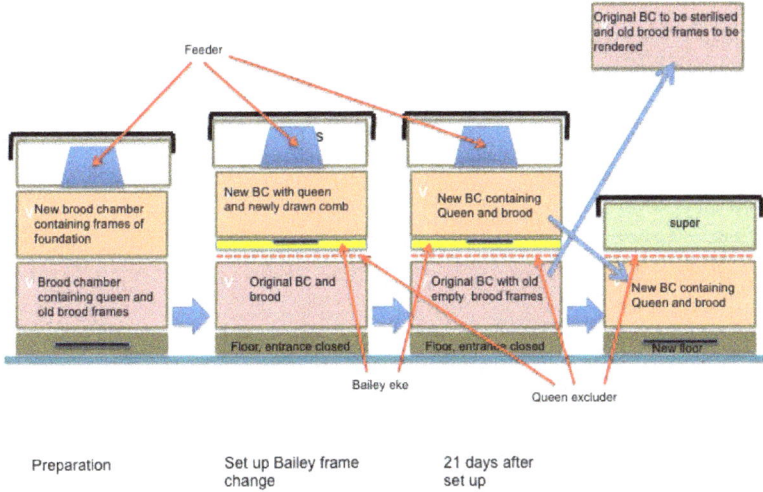

Original BC to be sterilised and old brood frames to be rendered

Feeder

S

New brood chamber containing frames of foundation

New BC with queen and newly drawn comb

New BC containing Queen and brood

super

Brood chamber containing queen and old brood frames

Original BC and brood

Original BC with old empty brood frames

New BC containing Queen and brood

Floor, entrance closed

Floor, entrance closed

New floor

Bailey eke

Queen excluder

Preparation

Set up Bailey frame change

21 days after set up

The configuration is designed to prevent the colony using the lower brood box to store pollen and nectar. It generally works well. In many cases when the queen is separated from the brood by a queen excluder, the colony will raise emergency queen cells. In my experience this does not happen in the lower box when a Bailey Frame Change is carried out. There are a number of possible reasons why this is the case. It could be that it is too early in the season for the swarming instinct to kick in. Alternatively, the configuration of the Bailey Frame Change, isolating the lower box from the main thoroughfare of the hive, could inhibit the workers' instinct to build queen cells there.

If the colony was overwintered with a super this can create a small complication. Before starting the procedure, any bees on the supers should be shaken into the brood box and then the supers can be put to one side until the feeding associated with the procedures is completed, at which time, they can be replaced on the hive above a queen excluder.

Though frame change is widely recommended, I'm not sure that the majority of beekeepers are carrying it out on a regular basis. But my experience over the past decade has convinced me that the effort is worthwhile, resulting in healthy and productive colonies that are easier to manage and less likely to swarm.

Chapter 13

Honey and wax

Your honey is like a mother or father's day card. It is all the better if it arrives when there is no expectation. If you keep your bees healthy then you have a chance of the bees producing honey, sometimes in embarrassing quantities. It's possible, not certain. The amount of honey you produce is not a judgement on your skill as a beekeeper. The main reason why experienced beekeepers get larger crops of honey is that, over time, they have selected apiaries where their bees do better. It is humbling but true that there are many factors that govern the amount of honey your bees produce other than your skill as a beekeeper, the most significant of these being the topology and flora of the area where your bees are based and the weather.

Honey

Honey is a marvellous substance. Since antiquity people have known that it is a healthy food. If you take that spoonful of honey every day you will live to be a hundred and fifty! I hope not! But the best cure I know for a cough and cold is hot honey and lemon. It can be used to improve your complexion, and its antiseptic properties mean it can be used for cuts and ulcers and during the first world war, before penicillin, honey was used to dress wounds.

It is, to a large degree, just a supersaturated solution of sugars. The water content varies between 17% and 20% (rather more for heather honey). The high concentration of sugar kills yeast, so whereas sugar solutions will normally begin to ferment, honey is stable for long periods provided it is sealed from direct contact with the atmosphere. The reason honey needs to be sealed from the atmosphere is because it is hygroscopic which means that it will attract and absorb water from the atmosphere, provided the air is sufficiently humid, and so if honey is left exposed, in time, the water content will increase to the point where it will support fermentation. But honeybees, which always have the answer, seal their honey in cells, capped with wax, isolating it from the atmosphere.

Approximately 77% of honey is sugar, 18% is water and the remainder is pollen and various aromatics which give the honey its distinctive flavour and smell. The sugars are mainly monosaccharides, fructose and glucose. The relative proportions of these sugars define the sweetness and texture of the honey. Fructose tastes sweeter, and is slower to granulate. Glucose is less sweet but granulates quickly. All honeys will granulate, given time. The slower a honey granulates the larger are the crystals. Honey from borage has a high proportion of fructose and so is a sweet honey that remains liquid for a long time. On the other hand, honey from oil seed rape nectar has a higher proportion of glucose, and so it readily crystallises with a fine crystal structure, producing a set honey, either hard set or soft set.

Above 30°C and below 8°C honey will not crystallise. Heat will convert a crystallised honey back to liquid and warming honey reduces its viscosity. This property is used when processing honey. But heat must be applied with care. If honey is heated for too long and at too high a temperature then HMF (hydroxyfurfuraldehyde) is produced. The honey regulations state that the maximum value for HMF in honey is 40mg per kilogram.

Nectar

The honey is mainly produced from the nectar that the bees collect. Nectar is a sugary solution produced by flowering plants that require insects for pollination. Flowering plants and pollinating insects evolved at the same time, each benefitting the other. The plants required a vehicle to carry pollen from one flower to another and to attract this vehicle, that is the pollinating insect, the flowering plants evolved with colourful flowers, aromas and nectaries in the base of the flower. The flowers have colourful and distinctive patterns which direct the insect to the site of the nectary. The patterns may include colours in the ultra violet range of the light spectrum, outside the human visual range but in the range of vision of the eye of the honeybee. The flower is designed so that accessing the nectary causes the bee to come into contact with the stamen carrying the pollen, which adheres to the hairs on the bees' body. Bees may visit many flowers before they achieve a full load of nectar, about 30mg. They tend to restrict themselves to a single species of flower so that the pollen that they have collected from one flower may come into contact with the female stigma of another flower of the same species and so bring about the pollination that the plants need. Although nectaries are usually positioned deep within the structure of a flower, some plants, such as broad beans, have nectar producing organs (extra floral nectaries) on their stems or leaves.

Not all honey comes from the nectar of flowering plants. Honeybees will also collect the sweet secretions of aphids on the leaves and twigs of trees to produce honeydew honey.

The properties of the nectar that honeybees collect varies according to the type of plant, the number of hours of daylight, the intensity of sunlight, the ambient temperature, humidity, the moisture in the soil, and the nature of the soil. These factors affect the quantity, the water content and the relative proportions of the different sugar types of the nectar produced. The predominant sugar in nectar is sucrose, a disaccharide, but there are also significant quantities of monosaccharides, trisaccharides, proteins and various aromatics

Converting Nectar to Honey

There are two main steps involved in the process that the honeybees employ to convert the nectar into honey. In the first of these, and this starts as soon as the nectar is taken into the honey stomach of the forager, the bee adds the enzyme invertase, produced in the hypopharyngeal gland situated in the head of the worker, to the nectar and this causes each molecule of the disaccharide sucrose to break down and combine with a single molecule of water to form a molecule of glucose and a molecule of fructose. A side effect of this reaction is to reduce the water content of the nectar. The second step of the process is to further decrease the water content to between 17% and 20%. This is achieved by the bees within the hive continually regurgitating and swallowing the nectar, forming a film of nectar at the entrance to their mouth, and then spreading the nectar thinly on the honey comb. The temperature of the hive is kept between 30°C and 35°C, and the bees ventilate the hive by fanning their wings in order to reduce the humidity and aid evaporation. In the quiet of a late summer evening the hives hum with the sound of the fanning as the colony processes the day's harvest and the air in the apiary is heady with the smell of nectar. Eventually the honey is ready, the cells of the honeycomb are full and the honey will be sealed within the cell with a cap of wax.

Processing the Honey Crop

Though it is not absolutely necessary it is usually easier and more efficient for the hobby beekeeper to remove the supers and then extract the honey as a single operation. Fresh honey from the hive extracts more easily and by extracting immediately it means that there is no necessity to store, and double handle the supers. Of course, if you have many hives then there may be no choice but to store your full supers to be extracted at a later stage. Prior to removing the supers, the bees need to be cleared from the supers. There are a number of methods of clearing bees from supers.

A honey comb partially capped

Installing a clearer board with two Porter bee escapes

Clearer Boards

These can be dedicated bits of kit but are usually adapted crown boards with a bee escape added as required. There are a number of designs of bee escapes but they are all designed to achieve the same end by acting as a valve that allows the bees to pass from the supers down towards the brood area, but does not allow or make it easy for them to move in the opposite direction. The most frequently used is the Porter bee escape. They are cheap, and are easily slotted into the oval hole that is cut into the top of the crown boards. Within the escape there is a pair of springs, which the bees can push aside as they move downwards from the honey super to the hive below. The springs can get obstructed with propolis and so, before being used, need to be examined, cleaned and adjusted if necessary, so that there is a gap of 5mm. It is possible to a use a pair of Porter escapes in each crown board. Whether one or two Porter bee escapes are used in each clearer board the supers can be cleared of bees in a day provided the escapes are clean and in good condition. Generally, it is not a good idea to leave full supers over an escape longer than is necessary. The supers full of honey are a tempting target and should there be an alternative entrance, through a hole in the woodwork, through a gap between the boxes or through the roof, the bees and any wasps will find it and empty the honey in a matter of days. Also, by leaving the Porter bee escape in place it will inevitably be clogged up with propolis. A crown board is not the place to permanently store the Porter escapes.

There are alternative forms of escape. The main ones are the rhombus escape and the hexagonal 6-way escape. They both are designed for rapid clearing on the principle that the route out of the super is easier for the bees to negotiate that the route back up into the supers. Though they clear the bees quickly, if left in place the bees will work out how to get back to the supers. In all cases it can aid clearing if a space is created above the escapes, either by using a clearer board with wider battens on the upper side or by inserting an eke above the clearer board.

Shake and Brush

In this method frames are removed one at a time from the super, the bees shaken off into the brood box and any that are still adhering to the comb are brushed off. The frame is then placed in an empty super and covered with a cloth. Not surprisingly this causes the bees significant upset and certainly should be avoided if you have close neighbours. Nevertheless, there are advantages. First, it requires only one visit to the apiary and second, it is the only approach if taking uncapped oil seed rape honey, when it is necessary to check each comb individually to ensure that the honey is ripe.

Blower

Petrol engine powered garden blowers are now relatively cheap and provide a quick and effective method of clearing bees from supers. The supers to be cleared are lifted off the hive and put on their side on the ground about 10m from the hive. Then using the blower, the bees can be blown from the combs. The bees will return to their home hive. Again, it is not a method to use where there are close neighbours.

Handling the Supers

Supers, which are full of honey, are heavy. National supers weigh about 14kg (30lb) and lifting them from the hive can put considerable stress on one's back. More elderly beekeepers should seek assistance in lifting full supers off the hive. Elderly beekeepers who struggle to lift a super may choose to remove the full frames of honey individually rather than in a full super. After removing supers of honey, care should be taken to prevent the bees having access. The supers should be stacked on a solid board with a second board on top of the stack. The joins between the supers should be checked to ensure that there are no gaps and if gaps do exist these should be sealed with duct tape. To put it mildly, it can be very disappointing to return to a stack of honey supers after a few days to find that they have been cleared of their honey.

Food Hygiene

The area that you use for processing honey may or may not be subject to food hygiene regulations. If you use the premises to process honey less than once a week and the honey that you produce is sold to consumers under your own name then the premises do not need to be registered for inspection. Therefore, most hobby beekeepers can use their own domestic kitchen without the intrusion of the authorities. Problems can arise if you produce honey that is sold in bulk to a packer. Even if you are not subject to inspection you are duty bound to make sure that the conditions under which you process the honey and the equipment used are spotlessly clean. Extracting honey is a sticky business and I take two important steps before starting, spreading out a plastic sheet on the floor and doing my best to persuade my spouse to vacate the home for several hours.

There are three main ways of extracting honey from the frames - using a centrifuge extractor, warming and pressing.

It is important that the water content of honey extracted is below 20%, preferably below 19%. It is possible that in selecting a super that is ready for extraction, the super will contain some frames where the honey is not fully ripe. This is particularly likely when extracting oil seed rape honey, which needs processing before it crystallises. A solution, which is not open to all, is to stack the supers with spacing between the boxes, in a sealed room, such as a conservatory, for 24 to 48 hours and have a dehumidifier running. This will reduce the water content to some extent.

Centrifuge Extractors

Centrifuge extractors come in many sizes. They may be made of food grade plastic or stainless steel (which is more expensive but easier to clean), be radial or tangential in configuration and be manually operated or powered by an electric motor. Radial extractors have a cage that supports the frames like the spokes of a bicycle wheel. Tangential extractors support the frames parallel to the circumference of the frame. Each has advantages and disadvantages. Tangential extractors can hold both shallow and deep frames and they will extract the maximum amount of honey, but the frames need to be switched about half way through the operation in order to extract the honey from both sides and the machine holds a smaller number of frames than a radial extractor of similar size. Radial extractors are not so effective at extracting honey, normally will not take deep frames but do not require double handling of the frames and take a larger number of frames. The most popular choice for the hobby beekeeper is the nine frame radial extractor, either plastic or stainless steel, manual or electric depending on the size of one's purse or biceps.

Extracting by Warming

To extract honey by warming it is usual to use a Pratley tray, also known as a melting tray or uncapping tray. Pratley trays are made from stainless steel and consist of a sloping tray above a bath of water which is heated by an electric element. Preferably the water temperature is controlled by a thermostat. Comb placed on the tray will melt and the honey flow down the sloping surface and through the outlet to be collected in a bucket. A Pratley tray can be used simply as a decapping tray. Care needs to be taken with their use. Excessive use of heat can alter the honey, driving off the aromatics and increasing the amount of HMF in the honey. The resultant honey is often of a darker colour than it would be if extracted by other means. Of course, if the honey has granulated in the frames, as is likely to happen with oil seed rape honey, there is simply no alternative but to melt it.

Extraction by Pressing

Honey presses can be purchased from beekeeping equipment suppliers. They are similar in construction to cider presses. The honey comb, having been cut from the frame, is placed in a linen bag which is then placed in the cylinder of the press which is made from perforated stainless steel. A piston is screwed down to compress the comb and the honey flows into a collecting bucket below the cylinder. The process is slower than using a centrifuge but is the usual method of extracting heather honey, which, because of its thixotropic properties, does not extract well in a centrifuge.

Honey Extraction in a Centrifuge

The processing of frames of honey in a centrifuge generally has the following steps

1. Decap the honey combs
2. Place the uncapped frames in a rack to drain
3. Place in the extractor and run until the honey is all extracted.
4. Drain the honey into clean 14kg plastic tubs for long term storage

Generally, it is advisable to separate the extraction process from the preparation of your honey for sale. The process of putting the honey in jars differs depending on whether liquid or set honey is being produced and this in itself will largely depend upon the type of honey the bees have given you. There is a demand for both, but my experience is that there is more demand for liquid honey than set honey. Fortunately, my bees produce more honey suitable for being sold as liquid rather than as set honey.

Honey regulations

There are regulations that must be complied with when preparing honey for sale. These are all set out within the food standards agency website. The areas addressed are the description of the honey, the information that must be included on the label, the layout of the label and the limits on some contents within honey such as hydroxyfurfuraldehyde. The major part of the regulations were formulated in 2003, but they are amended over time, most recently in 2015. For instance, the limitations on the size of jars have been recently removed. The important point is that the beekeeper is aware of such regulations and when designing labels ensures that the labels conform to the regulations.

Liquid Honey

The main summer crop, for me, comes from the nectar obtained from blackberry, lime, rosebay willow herb and balsam and which the bees blend to make a characterful and delicious liquid honey. It is not a unifloral honey but I'm more than content to accept the blends that the bees produce for me. After extraction the honey is initially stored in 14kg buckets and when required for sale is warmed in a warming cabinet, at approximately 40°C for 24 - 36 hours. The warming cabinet is an essential piece of equipment. They can be purchased but also they can be made as a DIY project, provided you have the necessary knowledge and skills. The heating element, which may be a 60W bulb, is connected in parallel to a fan and in serial with a thermostat. The walls, base and roof of the cabinet need to be well insulated, and there is a multiplicity of building materials that can be used. Refrigerators can be converted into warming cabinets. The other essential is a stainless steel settling tank with a coarse filter to remove fragments of wax and a fine 200 micron nylon cone filter that can be suspended within the settling tank, enabling the coarse and fine filtering to be completed as a single operation.

After warming, the honey is filtered and then allowed to settle for several hours before being put into clean, dry jars. The jars can be bought in bulk, often through the beekeeping associations. The standard honey jar in this country is the one pound squat jar, but many beekeepers now prefer to use hexagonal 8oz or 12oz jars. Before use the jars and lids should be washed, rinsed and stacked upside down to dry. It is not advisable to dry the jars with a cloth as this can leave small flecks of material on the glass. Before filling the jars a little experimentation is required to find the correct level to fill the jars to. The settling tanks are fitted with a honey tap which cuts off the honey flow quickly and cleanly.

After screwing on the lid, the jar can be sealed with a tamper proof seal, which can be an opportunity to advertise that you belong to a beekeeping association. And then the label needs to be stuck on. Most beekeeping suppliers will supply labels, customised with your own name and lot number. Buying labels in this way at least will ensure that they comply with honey regulations. The choice of an attractive and well designed label can certainly influence potential customers to buy your honey.

Soft Set Honey

For those of us who keep bees in arable farming areas where oil seed rape is a major crop, much of our honey will readily crystallise. If oil seed rape honey is poured directly into jars it will set firm, to the point of being knife bending. A better, indeed a very acceptable solution, is to sell the honey as soft set. Immediately after extraction the honey should be warmed, filtered, and then left to crystallise in 10 litre buckets (about 30 pounds). Provided the honey has a low water content it should be able to stay in this form indefinitely.

When it is required to be prepared for sale the bucket and the 30lb of honey can be warmed in a warming cabinet. The temperature required is about 40°C - 45°C and the time in the warming cabinet required is about a day. The honey at this point should be almost liquid, but there will be still some crystallised honey in it. The honey can then be poured into a stainless steel settling tank and periodically mixed with a creaming tool over a period of 24 hours or more until the honey becomes creamy, stiff and has a homogeneous consistency. At this point the honey can be put in jars. The honey will continue to set. The quicker the set the better, as this will give a finer consistency, and to achieve this, the jars of honey should be kept at 14°C – 15°C.

There is an alternative approach if you have samples of good quality soft set honey, with a smooth texture, which can be used to seed the crystallisation. The honey must be first heated until it has liquidised totally and then poured through the filters into the settling tank. Allow the honey to cool to about 25°C before adding the seed honey, adding about 10% by volume. Then, as before, periodically mix with a creaming tool over a period of 24 hours or more until the honey becomes creamy, stiff and has a homogeneous consistency and then bottle.

Comb Honey

Selling honey in the comb was the traditional method of trading honey before glass jars became a low cost item. Many older people remember as children being given comb honey to eat and so there remains a small but persistent demand for it. And for honey aficionados it remains the best way to obtain and eat honey, and for good reason. Honey in the comb is pure and unprocessed, just as the bees prepared it. It has not been filtered or heated. Whether the consumer wishes to eat the wax of the comb is a personal choice. The human system is unable to digest the wax. Comb honey can be spoiled by the presence of wax moths or braula coeca, the bee louse, and the honey comb needs to be checked carefully before being used as cut comb. The premium product at any honey show is heather comb honey. Comb honey will keep indefinitely if stored in a freezer.

Comb honey can be produced either as cut comb or in sections. Cut comb, as the name applies, is cut as rectangles from honey comb which is not wired. Comb Honey can be produced by using thin non-wired foundation or starter strips of foundation attached to the upper edge of the frame by the wedge. The small number of beekeepers who persevere or experiment with top bar beekeeping will primarily produce comb honey. The comb needs to be evenly built up, totally capped on both sides and removed from the hive shortly after being capped, so that it does not become discoloured with the marks of thousands of minute feet walking across it. Cutters are available to cut out rectangles of comb or a sharp warm knife will do equally well. It is possible to buy plastic containers with transparent lids designed to hold the cut comb. It is necessary to individually weigh and price each piece.

Section comb honey is produced in special racks, which are used in place of a super, in which are set a matrix of wooden (or plastic) sections, either round or square, in which the bees are encouraged to build their honey comb. It requires a strong colony, a good flow of nectar and an experienced and skilled beekeeper. The comb is sold in the sections in which the bees built the comb.

Wax

At one time wax was equally valuable as a product of the hive as honey. Certainly, the honeybees kept by monasteries were primarily used to produce beeswax for the candles that would burn in the abbey church. Of course, the odd glass of mead, no more than an unintended by-product, didn't go amiss either. Beeswax candles burn with a bright and pure flame and this, along with the faint sweet smell of the wax, helped to invoke a feeling of spirituality in those holy places. The small number of abbeys in this country are still centres of beekeeping. For many years Brother Adam at Buckfast Abbey led one of the country's most important honeybee breeding programmes. I remember with pleasure a visit to Pluscarden Abbey near Nairn and speaking to the brother who managed the bees there. In modern times a beeswax candle adds a touch of panache to a suburban dinner party.

Beeswax is also used to manufacture polishes and cosmetic products such as hand cream and lip gloss. Though most do not, every household should have a block of wax in the kitchen drawer to lubricate sticking drawers and windows and wax threads before sewing on buttons. At one time woodworkers and needlewomen would always have a block of wax at hand. Archers use wax on their bow strings.

The honeybees produce wax from four pairs of glands, situated between the segments on the ventral side of the abdomen. The production of wax mainly occurs during the second week after the worker has emerged. The worker must ingest ample quantities of both honey and pollen to produce wax. The bees use about 8kg of honey to produce 1kg of wax. The carotenoid pigments from the pollen can colour the wax. Beeswax is a complex mixture of hundreds of carbohydrates and esters.

The important thing is its physical properties. At normal atmospheric temperatures it is a stable and fairly inert substance with considerable mechanical strength. It retains this strength up to a temperature of about 35°C but as the temperature increases above this point it becomes progressively more pliable. At about 63°C beeswax melts and at about 204°C beeswax has a flashpoint at which the wax can spontaneously ignite, which can be very dangerous. For this reason, beeswax should never be heated directly in a pan on a stove, but rather over a water bath in a bain marie, which will ensure that the temperature of the wax never exceeds the boiling point of water, though to prevent discolouration the temperature should be kept below 85°C. Its specific gravity is about 0.96 so it floats on water.

The beekeeper will obtain wax from three main sources. From cappings during the extraction of honey, from the renewal of comb and the cleaning of brace and burr comb from the frames as the hives are inspected during the summer months. When wax is produced by the workers it is usually white, but it can quickly become discoloured, mainly by coming into contact with propolis. Comb that is repeatedly used for rearing brood becomes dark brown, due to the bees' practice of sterilising the inner surface of the cells with a thin layer of propolis between each cycle of brood. In the course of a summer season the cells at the heart of the brood nest will be used for more than eight batches of brood.

Most beekeepers process their wax so that it can be traded-in with the beekeeping equipment suppliers. They will accept blocks that come directly from the collection tray of a solar wax extractor. The suppliers use the wax to make foundation. The manufacturing process begins with cleaning and filtering and so there is no requirement to filter the wax before offering it for trade-in, but it should be in a block and free of large amounts of impurities and honey.

The solar wax extractor is a useful piece of equipment and can be largely made in a home workshop. It consists of a sloping galvanised tray, with a course filter at the lower side, held within an insulated box, with a double- glazed upper surface. The wax is melted by the heat generated by the sun's rays. The molten wax flows down the sloping surface of the tray, through the course filter and into a collection container. The solar extractor needs to be positioned so that it faces due south. When there is clear sky in the middle of the summer a tray full of comb will melt and the wax flow into the container in a couple of hours. From the end of August, the sun is no longer high enough in the sky to provide sufficient energy. In my solar extractor, at any one time, it is possible to process up to four brood frames which have old comb that needs to be renewed, and the wax will flow from them leaving behind just a propolis skeleton of the comb. At the same time as extracting the old wax I am sterilising the frame. Colleagues have expressed concerns that the heat denatures the timber of the frame. If wax is required for competition then the best wax to use is that obtained from cappings. It should be processed separately from other wax and not heated excessively.

Wax for candles, polishes or cosmetics must be cleaned further. I melt the wax from the solar extractor in a bain marie and then filter it through milk filters that can be obtained from agricultural suppliers. A large variety of candle moulds are available and these result in products that can be sold at craft fairs. I make no claims to produce wax to prize winning standards.

Eileen's recipe for furniture cream

2.5 oz beeswax

0.5 oz pure soap

0.25 pt water

0.5 pt pure turpentine (not substitute)

Shred wax into jar and add turpentine

Shred soap into another jar and add warm water

Stand both jars in warm room all night

When both the soap and the bees wax are dissolved, stir contents of one jar into another

Store in a jar with a screw cap

Pollen

Pollen is collected by the bees to provide the colony with a source of protein, fats, vitamins and trace elements. Pollen is required to feed the larvae and produce wax. The house bees, the worker honeybees in their first three weeks of life, eat the pollen. This results in brood food and royal jelly being produced by the hypopharyngeal and mandibular glands, which are situated in the head of the bee and at a later stage in the production of wax by the wax glands in the abdomen.

The bees store pollen, but generally only sufficient to allow the colony to meet its immediate needs. When it is brought into the colony it is mixed with nectar and put into a group of cells at the periphery of the brood nest. The effort that the colony expends on collecting pollen depends upon the requirement for pollen that is being generated. As the bees are unable to store pollen for long periods without it spoiling, if no pollen is available in the surrounding area at any given time then the raising of brood and the building of comb will soon cease. However, the worker bees that survive the winter have internal stores of fats and protein which allow the colony to start raising a limited amount of brood in the early spring.

Pollen is a remarkably nutritious food. Within it are contained all the twelve essential amino acids which are required by animals, vitamins, especially those in the B category, trace mineral elements and lipids. But the bees recognise that no one pollen type provides a balanced diet and even when there are ample quantities of one type of pollen the bees will seek out other sources to supplement their diet. The nutritional properties of pollen have led to its use as a food supplement for humans. There is no doubt that pollen is extremely nutritious and can be useful if in some way your normal diet is likely to be deficient in some nutrients, which may be the case for vegans. I collect pollen in the spring and add it throughout the year to my morning muesli, giving it a nuttier tang. I have no reason to believe that my diet is in any way deficient, and therefore it is no more than being faddish, not that there is anything wrong with that. So, for this reason I collect pollen. It is a product for which there is small demand from the general public and which can be used by the beekeeper to give supplementary feeding to the bees in the early spring if there are no sources of fresh pollen available.

To collect pollen I've built a couple of special floors which incorporate a pollen mesh which the bees must pass through when they return from foraging. It causes the pollen pellets to be knocked from their corbicula, the pollen basket on their back legs. The pollen pellet falls through a mesh into a tray for collection. The tray needs to be emptied every couple of days and the pollen dried and cleaned before it is stored in the freezer, where it will keep indefinitely. It is possible to obtain electrically powered pollen drying trays, but the pollen can be dried satisfactorily by spreading on a dinner plate for a few days in a warm kitchen. Cleaning the pollen takes time and care as it can contain the odd bee leg or wing.

A Honey Show

A stall run by the beekeeping association at a local craft festival

Honey Shows

I occasionally enter honey shows, but without any conviction or success. It needs an obsession with detail and a competitive edge that I do not possess. But I acknowledge the value of the shows to encourage high standards in our production of honey and wax for sale and they are a shop window for beekeepers to present their craft to the general public, who can be amazed at the variety of honeys that are produced and the other items such as candles, mead and honey cakes that can be made from the products of the hive.

Last Word

There is much more that can be written on the keeping of bees, but not by me, at least for the time being. Good luck, dear reader. I'm off to stroll down my garden in our beautiful Eden in Herefordshire to look at my bees.

July 2019

www.ingramcontent.com/pod-product-compliance
Lightning Source LLC
Chambersburg PA
CBHW061217270326
41926CB00028B/4672